Astrophysics and Space Science Library

More information about this series at
http://www.springer.com/series/5664

Giants of Eclipse: The ζ Aurigae Stars and Other Binary Systems

Thomas B. Ake • Elizabeth Griffin
Editors

Giants of Eclipse: The ζ Aurigae Stars and Other Binary Systems

Springer

Editors
Thomas B. Ake
Computer Sciences Corporation
Space Telescope Science Institute
Baltimore
Maryland
USA

Elizabeth Griffin
National Research Council of Canada
Dominion Astrophysical Observatory
Victoria
British Columbia
Canada

ISSN 0067-0057 ISSN 2214-7985 (electronic)
ISBN 978-3-319-09197-6 ISBN 978-3-319-09198-3 (eBook)
DOI 10.1007/978-3-319-09198-3
Springer Cham Heidelberg New York Dordrecht London

Library of Congress Control Number: 2014958290

Cover illustration: Atmospheric-eclipse phases of a ζ Aurigae system. The small hot secondary is shining through different layers of the chromosphere of its cool-giant primary just before (or just after) phases of total eclipse.

Printed on acid-free paper

Springer is part of Springer Science+Business Media (www.springer.com)

Dedicated posthumously to the leadership, perseverance and scientific insight of K.O. Wright, one-time Director of the Dominion Astrophysical Observatory, and himself a giant in eclipse spectroscopy of the ζ Aurigae stars

Foreword

Although binary and multiple star systems make up about 60–70% of stars in the Galaxy, only about 0.5% are in systems that undergo eclipses; the requirement that the plane of a binary's orbit be aligned edge-on is sufficiently stringent (the more so when stellar separations are large) that such alignments are very unlikely. This is a serious limitation because eclipsing binaries play crucial roles in astrophysics by providing, among other things, fundamental measurements of stellar masses, radii, and luminosities. But among those rare eclipsing binaries are a few even rarer ones that offer crucial information on the structure and physical conditions of stellar atmospheres, interiors, and evolution that currently cannot be secured by any other means. In the case of the ζ Aurigae binaries (the subgroup highlighted here), the small, hot component serves as a beaming probe of the outer atmosphere, or chromosphere, of its much larger cool giant or supergiant companion just before and just after being eclipsed by that star. The favorable geometry offers a powerful laboratory that allows the physical properties of the cool star's chromosphere to be determined at varying distances from the stellar surface.

The primary source of information about these fascinating objects has for nearly half a century remained the review by K.O. Wright in *Vistas in Astronomy*, 1970. My own copy of that paper is well worn and dog-eared from persistent use over the years. It is a well-deserved testament; even though the interferometry, UV spectroscopy, sophisticated theory and modelling of stellar chromospheres, and asteroseismology of modern research were not available, or not at all developed, in Wright's time, it is amazing to appreciate how much had nevertheless been learned about the ζ Aurigae binaries from early photographic spectroscopy and single-cell photoelectric photometry. All the same, the introduction of CCD spectroscopy in the 1980s and the advent and subsequent burgeoning of space-based data, especially in the ultraviolet, suggest a need to update Wright's review with the new results from those advances, and "Giants of Eclipse" (which is in fact appropriately dedicated to K.O. Wright) will now serve as that new reference.

"Giants of Eclipse" builds on the results derived from classical visible-wavelength spectroscopy by engaging the benefits of high-signal-to-noise CCD observations and then weaves in the additional data that have accrued from space missions and from the developing technologies of interferometry and asteroseismology, while at the same time maintaining a focus on the complementarity of that research to other important areas of astrophysics. That dual role will ensure the value of the book as a source of astrophysical information for many years to come.

The seven chapters that comprise "Giants of Eclipse," though written by a consortium of 12 experts, have been integrated almost seamlessly into a whole that offers enjoyable reading both for those with a general interest in astronomy and also for specialists, professionals, and students of eclipsing binaries, evolved cool stars, stellar atmospheres, and stellar evolution.

Villanova, USA Edward F. Guinan

Acknowledgements

To every book there is a unique set of complexities. The ones that beset us in the present instance derived from the complexity of the topic itself. Although the 1970 publication (in a refereed journal) that we planned to update had a single author (Dr. K. O. Wright), both the subject material and the available methods of attack had evolved and burgeoned sufficiently since then that we needed to call on many different sources of expertise—10, to be precise—plus 2 others whose contributions, while relatively short, were as profound in their own areas as the rest. It is to those 10 co-authors and contributors that we owe our most sincere gratitude, not only for their expert compositions but also for allowing us to demand—and have—their time and attention when it was convenient to us. Each leads a busy academic life, yet each proved capable of those extra achievements which (so the adage attests) busy people do best. To them we offer our thanks for fine science and thorough work, for prompt co-operation and rapid responses, for flexibility and good nature, and (mirabile dictu) for remaining our friends.

For their part, several of those authors have asked to include an acknowledgement of individual funding support, and we do so hereunder:

Phil Bennett (Chap. 3) acknowledges that his research was made possible by the generous awards of Hubble Space Telescope observing time by STScI for GO programs 7269, 8257, 8779, and 9231. Robert Stencel (Chap. 4) acknowledges partial support from NSF grant AST10-16678 to the University of Denver, and from the William Herschel Womble bequest in support of astronomy at the University of Denver. In adding personal gratitude to colleagues, collaborators, and competitors he singles out Jeffry Hopkins and the AAVSO, and also Kathleen Geise and Richard Pearson of the University of Denver. Vladimir Airapetian (Chap. 5) has been supported by NASA grant NNG09EQ01C, and Manfred Cuntz (also Chap. 5) by Program number HST-GO-13019.02-A, provided by NASA through a grant from the Space Telescope Science Institute, which is operated by the Association of Universities for Research in Astronomy, Incorporated, under NASA contract NAS5-26555. Daniel Huber (Chap. 7) acknowledges financial support through an appointment to the NASA Postdoctoral Program at Ames Research Center administered by Oak Ridge Associated Universities, and NASA Grant NNX14AB92G issued through the Kepler Participating Scientist Program.

Tom Ake acknowledges financial support from the Computer Sciences Corporation (CSC) NPS Mission Services Science and Engineering group to attend the Monterey AASTCS meeting, and thanks are also due to the AAS for its significant financial support of that meeting. Finally, Elizabeth Griffin would like to acknowledge the privileges of the Volunteer Visitor programme which is operated at the Dominion Astrophysical Observatory (National Research Council of Canada), for generous awards of DAO telescope time, particularly during eclipse phases of most of the ζ Aur binaries, and for upgrading and maintaining the in-house PDS, with which numerous archived photographic spectrograms were digitized.

Contents

Contributing Authors

Vladimir S. Airapetian
Thomas B. Ake
Wendy Hagen Bauer
Philip D. Bennett
Manfred Cuntz
Joel A. Eaton
R. Elizabeth Griffin
Daniel Huber
Brian Kloppenborg
Klaus-Peter Schröder
Robert E. Stencel
Gerard van Belle

Chapter 1
The ζ Aurigae Binaries

R. Elizabeth Griffin and Thomas B. Ake

Abstract This opening chapter provides a brief historical overview of the ζ Aur stars, with a focus on what K.O. Wright, his predecessors and colleagues at the Dominion Astrophysical Observatory, and his contemporaries further afield, achieved during the era of pre-electronic data. It places the topic within the framework of modern observing, data management and computing, outlines the principal features of the chromospheric-eclipse phenomena which single out the ζ Aur binaries for special study, and describes the considerable potential which this remarkable yet very select group of stars offers for increasing our understanding of stellar physics.

1.1 Introducing the ζ Aurigae Binaries

The Galaxy contains over 10^{10} stars, and quite a number of them have at some time, or will soon be, classified in some manner, and placed in the 'boxes' that are the tools of the scientific statistician. Many will exhibit normality (however we define that); at least as many more will exhibit deviations from that normality, and the statistics will reveal that certain of the trends which can then be discerned are astrophysically significant. Yet it took only one single star in that entire Galactic sample to demonstrate a quite unanticipated but very remarkable phenomenon—an atmospheric eclipse—which opened up a route to new science that had not previously even been imagined. It allowed us to observe directly a star's chromosphere—something that could only be done in the case of the Sun—and thereby to study those outermost regions where a star interacts with its own immediate environment, and into which it sheds material from its surface. The astrophysics which developed from that one star

R.E. Griffin (✉)
Dominion Astrophysical Observatory, National Research Council of Canada, 5071 West Saanich Road, Victoria, BC, V9E 2E7, Canada
e-mail: Elizabeth.Griffin@nrc-cnrc.gc.ca

T.B. Ake
Space Telescope Science Institute/CSC, 3700 San Martin Drive, Baltimore, MD 21218, USA
e-mail: take@stsci.edu

T.B. Ake, E. Griffin (eds.), *Giants of Eclipse: The ζ Aurigae Stars and Other Binary Systems*, Astrophysics and Space Science Library 408,
DOI 10.1007/978-3-319-09198-3_1

burgeoned into a substantial contributor to stellar physics; observers and modellers cornered international photometric and spectroscopic observing campaigns during the twentieth century, collectively requested major amounts of observing time on telescopes in space as well as on the ground, and still continue to wrest all possible information from both archived and new data. Studies of this star and a small handful of related objects also commandeered an AASTCS conference, *Giants of Eclipse*, held in July 2013 in Monterey, California. The star is ζ Aurigae.

1.2 History and Background

The history of this special field of astrophysics dates back nearly 90 years. The newly-completed 72-in. reflector at the Dominion Astrophysical Observatory (DAO) near Victoria, Canada, was being used to refine the radial-velocity (RV) orbits of stars recently flagged as binaries by the extensive Lick RV surveys. ζ Aur had already been classified as a 'composite-spectrum binary', a term coined before 1900 by the Henry Draper classification team at Harvard to describe a binary in which a cool giant star is paired with a much hotter and comparatively unevolved star. Although the giant is considerably more luminous that the dwarf, their flux distributions have quite different shapes, and in the blue and near-UV spectral region they are fairly comparable. The observed 'composite' spectrum results from the superposition of the spectra of the two components. One noticeable characteristic is that the plentiful, deep absorption features that are typical of the spectrum of a late-type giant appear substantially weaker than in a similar single star owing to the veiling effect of the flux from the hotter dwarf companion. In ζ Aur, whose hot component is a mid-B dwarf, that veiling is very pronounced in the photographic blue spectral region. When Harper noticed that in one of his spectra of ζ Aur taken at the DAO in 1924 the contribution from the hot component seemed to be *absent*, he surmised that it could be in eclipse (Harper 1924). The orbit then to hand was sufficiently accurate, and the duration of an eclipse sufficiently long, that the next event, as predicted by Bottlinger (1926), could not be missed, although only one eclipse in three was well enough placed for all-night observing. In early 1932 Guthnick showed Harper's surmise to be correct (Guthnick 1934).

But that wasn't all. Spectra obtained by Guthnick during ingress into the January 1932 eclipse, and by Beer during the 1934 September one (Beer 1934), revealed deep, narrow absorption features that waxed rapidly in intensity as ingress approached, and waned equally rapidly during egress. Most of those extra features corresponded to the familiar ground-state lines that are so prolific in a late-type spectrum, but included additional sharp cores in the Balmer lines (see Fig. 1.1). Furthermore, the shorter the wavelength, the longer totality appeared to last. These dramatic changes caught the attention of theoreticians, and in 1936 Menzel (Harvard) offered the explanation: although both stars were visible since the observed spectrum was still composite, the hot star was acting like a light-probe and shining through progressively thicker layers of the outer atmosphere of the K-type

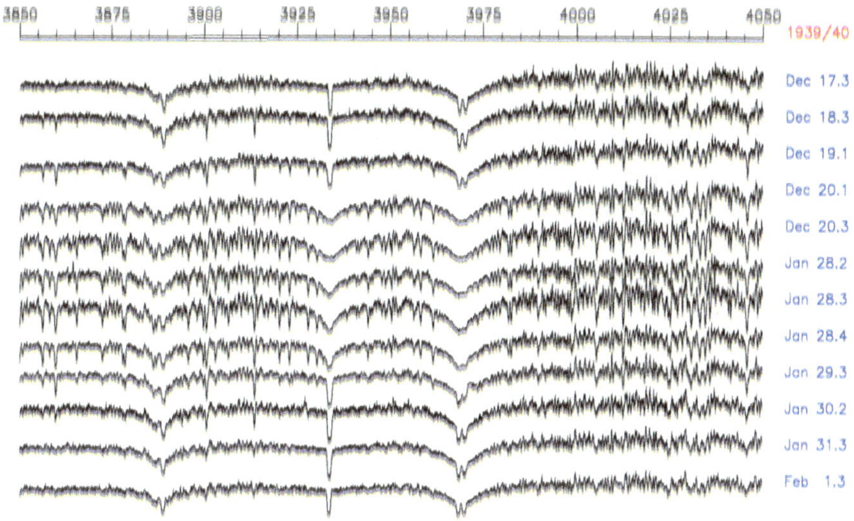

Fig. 1.1 The effects of chromospheric absorption observed during the eclipse of ζ Aur in 1939–1940. The original photographic spectra exposed at Mount Wilson were digitized at the DAO (see Appendix, Chapter 2, page 80). The spectra illustrated here are a representative selection of the full series. According to Kron (1940), first contact occurred on 1939 December 19 at 15:45 UT, with second contact ~1.7 days later, on December 21 at 08:30; totality lasted for nearly 37 days, with mid-eclipse on 1940 January 8. Third contact took place about 0:00 on 1940 January 27, and fourth contact ~1.7 days later

supergiant as it passed behind it towards first contact. In uncharacteristically graphic terms, Menzel likened the event to that of "a planet setting in a smoky atmosphere, disappearing before it reaches the horizon", its light as much extinguished as occulted (Menzel 1936). Figure 1.1 shows a sequence of spectra of ζ Aur observed near conjunction phases.

1.3 Chromospheric Eclipses

It was not difficult to explain Guthnick's observations *qualitatively*. As an eclipse approaches, the familiar composite spectrum, with its combination of broad early-type features mingled with numerous but weakened narrow lines from the cool primary, becomes complicated by the addition of a third spectrum, as the chromosphere selectively absorbs the hot star's radiation passing through it in the line of sight. Formed by pure absorption in a region of very low density, those chromospheric lines arise from atoms and ions in their lowest excitation states. They are typically very narrow and without wings, and the strongest ones can be almost black in the core as very little scattering takes place. As the light from the hot star traverses increasingly deep layers of the chromosphere, the narrow absorption lines strengthen noticeably until second contact, when the spectra of both the early-type star and the chromosphere vanish and only that of the cool primary remains. The reverse then happens from third contact onwards. A time-series of spectra like the one reproduced in Fig. 1.1 samples the column density of the chromosphere at different distances from the primary's limb, thereby providing unique information for modelling the density and temperature gradients in the neighbourhood of the secondary's projected trajectory—though constructing those models accurately can be an exacting challenge.

Calcium is one of the most abundant photospheric metals, and is readily observed in the blue or near-UV spectral region through its strong resonance lines of Ca II at λ 3968 and 3933 Å (Fraunhofer's H and K lines). The low temperature and rarefied conditions of a stellar chromosphere favour absorption from ground-state or low-excitation levels of once-ionized species, so the H and K features become very pronounced during a chromospheric (*a.k.a. atmospheric*) eclipse and effectively represent the chromosphere throughout much of its discernible extent. Other features, such as the four strong neutral iron lines between λ 3920 and 3930 Å, also show chromospheric absorption, but are only strong enough to be detected in the lowest, most dense, layers where the chromosphere merges with the 'upper photosphere'. In Fig. 1.1 the secondary star was partially eclipsed on December 20 and January 28—as verified by Kron's (1940) photometry—so the line of sight to it only sampled the lowest, most dense, chromospheric layers. The Ca II lines were so strong at those phases that they had by then developed pronounced damping wings.

The series illustrated in Fig. 1.1 was derived from one of several comparable sets of ζ Aur spectra recorded at Mount Wilson, mostly during the winter eclipses. A number of similar sets was obtained at the DAO (though poor observing conditions often interfered at that season), and also at several other observatories across the world. In the early years of ζ Aur research, monitoring the chromosphere was usually restricted to partial-eclipse phases and to the nights of deepest chromospheric absorption close to first and fourth contacts, when chromospheric lines are very plentiful. All the same, as Fig. 1.1 shows, the chromosphere of ζ Aur could still be detected at both the start and the finish of that series of observations. In fact, the giant's outer chromosphere extends to distances that are substantially greater than the radius of its photosphere, and can be detected spectroscopically for several weeks both before and after total eclipse. Faraggiana (1965) reported that, according to spectra of the 1963–1964 eclipse taken at Merate Observatory, the chromospheric K line first appeared about 38 days before first contact and vanished about 24 days after fourth contact.

1.4 The Field Widens

In 1950 two more systems, 31 Cyg (McLaughlin 1950b) and 32 Cyg (McLaughlin 1950a), were added to the collection of eclipsing composite-spectrum binaries, now collectively known as 'ζ Aurigae systems'. Though broadly similar to ζ Aur in nature, 31 Cyg and 32 Cyg present different characteristics of mass ratio, flux ratio and spectral types, as well as orbital period and inclination. The intensities and small-scale complexities of their chromospheric absorption lines also exhibited behaviour that was more capricious than in ζ Aur itself.

Well before 1950, the 'ζ Aur effect' had also been recognized by Gaposchkin (1937) in the M supergiant plus B dwarf which comprise the long-period binary VV Cep. The flux from the M star overwhelms that from the B star in the optical region, so the system is best studied in the UV. However, the changes which spectra of VV Cep manifest are greatly complicated by variable emission within the system, believed to arise mainly through the transfer of mass from the supergiant onto its companion, and by circumbinary and interstellar features, all of which are enhanced at UV wavelengths. The challenges and problems presented by VV Cep are sufficiently idiosyncratic that in this book they have been given a separate discussion (Chap. 3, page 85).

In 1978 *IUE* opened up spectroscopy at space-UV wavelengths, and the binary 22 Vul became the next new addition to the ζ Aur class when Parsons and Ake (1983) recognized not only the composite nature of its spectrum but also that it underwent an eclipse (Parsons and Ake 1984)—though only for about 8 days, unlike VV Cep in which an eclipse lasts for about a year and a half. As with VV Cep, the dominance of the flux from the supergiant primary (in this case of mid-G type) throughout the optical region explained why it had not been detected earlier from the ground.

The discovery of a total eclipse in HR 6902 in 1986 (Griffin 1988) was followed by the discovery of a partial one in HR 2554 (Ake and Parsons 1987). Suspected eclipses in the G8 III+A4 V binary τ Per, also confirmed as partial by *IUE* observations (Ake et al. 1986), were studied both spectroscopically and photometrically during its 1989 conjunction (Griffin et al. 1992). Recognition in 1990 that the relatively short-period composite-spectrum binary HD 223971 (G7 III+A2 V; $P = 50.1$ days) is an eclipsing system (Griffin et al. 1991) was itself eclipsed by the somewhat surprising discovery very shortly afterwards that γ Per (G7 III+A2 IV), a bright system ($V = 2.9$ mag.) with a period of 14.6 years, undergoes total eclipses that can easily be detected visually and last for more than a week (Griffin 1991b). These subsequent additions to the ζ Aur class contain G-type primaries of somewhat lower luminosity, and therefore of smaller physical size, than in the first five systems to be discovered, thus extending the scope of the information that can be extracted from a chromospheric eclipse but bringing new challenges in extracting it.

Along the way, one system was 'lost' from the class. δ Sge (M2 IIab + B9.5 V) had been reported by Hynek (1942) as exhibiting sharp, variable Ca II K-line cores suggestive of at least an atmospheric eclipse, and the system was forthwith dubbed a member, or probable member, of the class. Because the orbital period is 10 years it took time to re-observe it astutely with the right equipment before enough evidence was to hand to demonstrate that in fact its orbital inclination was a mere 35° or so rather than near the 90° which eclipses require (Griffin 1991a). The observed changes in the K-line core, which have since been examined at high dispersion, appear to be phase-related combinations of broad but weak absorption intrinsic to the late-B star, narrow interstellar absorption, and variable emission. Unlike the early-B star in VV Cep (described in Chap. 3), the B-type component in δ Sge is not energetic enough to produce its own emission spectrum—its photospheric spectrum is fully present and visible—so the emission is likely to be associated with the M-type primary. That emission may cause the photometric variability that has been widely reported (e.g., Demircan et al. 1990), and may be associated with low-amplitude periodicities in the radial-velocity residuals that have also been detected (Walker et al. 1995).

The positive identifications bring to 10 the number of known ζ Aur binaries. As a sample it is not large, and since no two cases are very similar there is little meaning in defining a norm. Some of the fundamental characteristics and descriptors of the systems have been assembled in Table 1.1. The values listed there are not new determinations, nor necessarily consensus ones, but have either been taken from various papers or have been recalculated and not yet published. The purpose of the Table is to show that the stars in question are all bright objects (all but one are visible to the unaided eye), that their periods range over two orders of magnitude, and that the eclipses which they exhibit are, except for the last two, rather difficult to discern by eye, though all except that of HR 2554 are more pronounced at shorter wavelengths. The range and types of chromospheric-eclipse phenomena which they encompass extend from the very dramatic to an apparent absence of visible information, though a null result is not without scientific significance. In fact the very diversity which they collectively exhibit speaks to the diversity of all cool-giant stars—their structure, evolutionary states, mass losses, and the extent to which small differences in the temperature or pressure gradient of the outer atmosphere can affect, or can be affected by, those parameters.

Yet while that diversity checks our ability to model *the* archetypal stellar chromosphere since in practice there seems to be no such entity, it demonstrates vividly the wide range of physical laws in play and brings home how wrong our *photospheric* models and assumed mass-loss mechanisms would be without that extra insight which the phenomena of atmospheric eclipses alone can provide. The reviews in the following chapters illustrate and discuss that richness and potential, and focus on observations made both from the ground and from space. When we reflect that it is only in the case of the Sun that a star's chromosphere can be observed directly, the true worth and exciting potential of these atmospheric eclipses begins to be appreciated. Finding ways to realize that potential is what this book is about.

1.5 Other Advantages Offered by the ζ Aur Stars

The ζ Aur systems not only provide unique information about stellar chromospheres, but they also occupy a special position within the wider context of stellar research. The mass of a star is generally considered to be its most fundamental parameter. The amount of mass in a star governs the rate at which it evolves, but

Table 1.1 Members of the class of ζ Aur Stars

System	Alternative designation	Spectral type	V (m_V) (mag.)	Orbital period (days)	e	Total eclipse (days)	Eclipse depth ΔV (mag.)	ΔB (mag.)	ΔU (mag.)	Note
ζ Aur	HD 32068/9	K4 Ib + B5 V	3.8	972	0.38	37	0.15	0.55	1.95	(1)
31 Cyg	V695 Cyg	K4 Ib + B3-4	3.8	3784	0.21	61	0.10	0.41	1.48	(2)
32 Cyg	V1488 Cyg	K5 Ib + B6 V	4.0	1148	0.30	Grazing	0.05	0.21	1.06	(3)
VV Cep	HD 208816	M2 Iab + B0-2	4.9	7430	0.35	233	0.05:	0.40	1.92	(4)
22 Vul	QS Vul	G7 Ib-II + B8.5 V	5.2	249	0.0	8	0.05	0.13	0.35	(5)
HR 6902	V2291 Oph	G9 IIb + B8.5 IV	5.7	385.0	0.31	3.8	0.19	0.44	1.06	(6)
HR 2554	V415 Car	G7 II + A1 V	4.4	195.3	0.0	Partial	0.06	0.08	0.16	(7)
τ Per	ADS 2022	G8 III + A4 V	4.0	1516	0.73	Partial	0.16	0.30	0.45	(8)
γ Per	ADS 2324	G8 II-III + A1 IV	2.9	5328	0.79	7.3	0.28	0.54	0.88	(9)
HD 223971	V413 And	G7 III + F2 IIIm	6.6	50.1	0.0	<1	0.41	0.64	0.85	(10)

Notes: (1) Bennett et al. (1996). (2) Eaton and Bell (1994). (3) Burnashev and Burnasheva (2011). (4) Saitō et al. (1980); photometry affected by intrinsic variability. (5) Griffin et al. (1993). (6) Griffin et al. (1995). (7) Brown et al. (2001). (8) Griffin et al. (1992). (9) Griffin (1991b). (10) Unpublished

unfortunately it is notoriously difficult to determine at all accurately the mass of a single star. If the star is a single-lined binary (SB1), its RV orbit yields the mass function of the unseen star, though that can give no more than a minimum value for the mass because the orbital inclination is usually unknown—unless supplied from an astrometric orbit. If the binary is double-lined (SB2), the orbit solution yields the ratio of the masses of the component stars, $q = M_1/M_2$, and if—as for the ζ Aur systems—those component stars are at very different evolutionary states, then comparing their observed positions in the H–R diagram with theoretical evolutionary tracks provides a stringent test of stellar evolution theory, as has been demonstrated effectively by (for example) Schröder et al. (1997). But if the binary system also eclipses, the individual masses can be obtained absolutely from the SB2 orbit.

The largest uncertainties in the methods just outlined are introduced by the RV measurements of the hot secondary, particularly if it is much hotter than A0 and is also rotating fairly rapidly. A B-type spectrum has few features—just hydrogen lines, which are intrinsically very broad, and a handful of rather weak lines of He I and metals—and rapid rotation will cause weak lines to be washed out and difficult to discern, let alone measure precisely. On the other hand, the greater density of features in A-type spectra may cause blending with lines of the late-type giant, though cross-correlation techniques for determining velocity shifts are usually able to cope well if that occurs.

Programmes to determine stellar masses from SB2 orbits have been pursued by many groups and with many instruments, both on the ground and in space. It was attempts to refine the mass ratios of HR 6902 and HD 223971 by this method that led to the discovery of their respective eclipses. The complementary knowledge gained from both aspects of that research has subsequently been mutually enriching.

1.6 Atmospheric Eclipses: Pioneers and Leaders

Monitoring the ζ Aur system through eclipses was pursued vigorously by the pioneers of ζ Aur research, the lead often being assumed by the DAO, where the interest initially sparked by Harper had—and still has—retained its original lustre. K.O. Wright, a long-time DAO staff member and later its Director, was a key figure in the indefatigable mining of the resources of the ζ Aur systems. The astrophysical nuggets which only those systems can yield was one of Wright's lifetime fascinations, and with the prerogative that only a Director can muster he would request his staff to record spectra of the three classical systems, and of VV Cep, whenever appropriate eclipse phases occurred and the objects were suitably placed for observing. Wright was well aware how the signatures of chromospheric absorption could, and frequently did, alter not only from eclipse to eclipse but also from night to night, revealing 'satellite' absorption components that made serendipitous velocity excursions over and above the anticipated steady changes correlated with the projected height of the secondary star from the supergiant's limb.

In tandem with DAO colleagues, principally McKellar, and with others further afield, Wright co-ordinated worldwide observing campaigns, making use of the International Astronomical Union's channels to communicate with those interested and able to contribute photometric or spectroscopic data. Since the partial eclipse phases of all three systems last for 1 or 2 days it is in any case necessary to obtain observations at all longitudes, though they need to be collated carefully in order to avoid systematic or zero-point differences between different sets. For ζ Aur, though, a winter-time object and having a period very close to $2\frac{2}{3}$ years, only one eclipse in three (the winter-time ones) can be observed all night from early ingress to late egress, so obtaining broad back-up coverage for such observations was also a wise investment. One of Wright's chief legacies has been the rich collection of high-dispersion photographic spectra of the three classical systems and of VV Cep that were observed at the DAO. Before 1962 all DAO spectra were obtained with the 72-in. telescope and Cassegrain spectrograph, but later ones were observed at higher dispersion with the 48-in. telescope and its dedicated coudé spectrograph, described in detail by Richardson (1968). With the DAO's accurate 'PDS' plate-scanning system it has now been possible to convert those observations into a digital FITS format for dissemination into the public domain. Although Wright and his colleagues milked those spectra fairly intensively at the time, they did not have the tools to manipulate them as can now be done, so there is much to be gained by re-working those historic series of data.

Wright was not only passionate about the new astrophysics which these eclipses could offer to anyone who worked hard enough at them, he was also a man before his time. He could visualize clearly the benefits of removing the spectrum of the cool giant from each composite one, and though restricted to working in analogue mode he devised a cunning (but tortuously slow) graphical differencing method for isolating the chromospheric spectrum (Wright and Lee 1956). Interpretations also entailed knowing the brightness ratio of the component stars at different wavelengths, a quantity that is still difficult to measure very precisely today. Nevertheless, he and his collaborators did succeed in recovering or de-blending quite creditable Balmer-line profiles of the spectrum of the secondary star in ζ Aur (Lee and Wright 1960), and of chromospheric lines in 31 Cyg (Wright and Odgers 1962) and 32 Cyg (Wright and Hesse 1968).

Wright also accumulated a comprehensive familiarity with published efforts to observe and interpret chromospheric-eclipse spectra, and in 1970 he prepared a major review of that information for *Vistas in Astronomy* (Wright 1970). He included ϵ Aurigae among the cast of stars that showed chromospheric eclipses, and while later research has demonstrated conclusively that the eclipse phenomena in ϵ Aur cannot in fact be explained in that way, we too have included ϵ Aur in this volume—but for a different reason, as explained below.

Since K.O. Wright's time, astronomical research has been enormously influenced by major technological advances, in particular the opportunities to make observations from space at wavelengths not reachable from the ground, and the introduction of electronic means to acquire and handle data. What those advances meant for stellar physics were computer techniques for calculating large grids of models of stellar atmospheres, merged and sorted on-line databases of atomic data for spectrum synthesis, free access to archived observations, and comprehensive on-line access to astronomy's literature. Astronomical spectra that were once recorded as unique physical objects could—in electronic form—be mined, transferred and researched by geographically separated communities in a flash, modelled with a pandora's box of packages, and the results communicated in an instant. At a more modest level, the ability to manipulate composite spectra and to separate their individual components has provided a very welcome means for uncovering and isolating chromospheric features accurately. Some of those opportunities are described and illustrated in Sect. 2.4. The benefits have been particularly noticeable in research into the ζ Aur stars.

1.7 Atmospheric Eclipses: Opportunities and Challenges

An eclipse of any kind is an awe-inspiring event, whether viewed directly or observed photometrically, but the spectroscopic changes that are associated with an atmospheric eclipse are surely among the most dramatic that the cosmos offers. What makes it so remarkable is not so much the nature of the changes—the growth or decline of certain absorption features—but the speed with which they can occur. Major changes in spectral type are of course known to occur throughout the course of stellar evolution, but the most rapid happen on timescales of one or more centuries whereas the changes that signal a chromospheric eclipse are noticeable in just 1 day, and sometimes less—as can be discerned in Fig. 1.1. While it is not rare in astrophysics for the light from one body to be intercepted or modified by a foreground one that is in the line of sight—one can mention exoplanets, foreground stars in fields of clusters or extended objects, foreground galaxies responsible for gravitational lensing, and countless eclipsing binaries—it is extremely unusual to be able to resolve the outer layers of an extended stellar atmosphere in the way that can be achieved by observations of an atmospheric eclipse.

The relative strengths of chromospheric features contain information about the temperature and density structure of the material in the line of sight to the hot star. The RVs of the features indicate how the chromospheric material is moving, whether rotating as a solid body (as if it were an extension of the photosphere), or whether independent clumps of material are moving randomly in both direction and speed. The observations which the early researchers analyzed produced convincing evidence that the chromospheres of stars like the classical ζ Aur systems contain a mixture of such structures: a regular one that is a quasi-extension of the photosphere and probably co-rotating with it, but intermingled with clumps that move independently and often much faster. By piecing together the results from analyses of different eclipses of ζ Aur, McKellar and Petrie (1952) concluded that its chromosphere is neither completely stable nor fully symmetrical, and although a single 'minimum' model could be constructed that described most of the density and temperature profiles during ingress and egress, enhanced concentrations of all the measured elements sometimes occurred, at either ingress or egress but rarely at both of the same eclipse. Wright and Hesse (1968) came to the same conclusion regarding 32 Cyg.

Many of the unique benefits of atmospheric-eclipse events in ζ Aur binaries derive from the favourable ratios of size and temperature of the component stars. In the classical ζ Aur systems the projected diameter of the B-type dwarf is only a few percent of the linear extent of the primary's chromosphere, and its relative orbital velocity will cause it to move through a distance comparable to the size of its own diameter in about 1 day. A nightly series of spectra will therefore sample the chromosphere almost continuously, and with good height resolution.

Unfortunately, there are problems to address. The inevitable close blending between the absorption lines of the chromosphere and those in the spectrum of the cool supergiant makes it difficult to separate confidently the chromospheric from the photospheric contributions to each line, and also tends to bias any RV measurements made during atmospheric-eclipse phases. That blending posed a major challenge to analyses based on non-digital data, as measurements had then to be made on spectra that were fully composite. Digital spectra provide a much-needed opportunity to isolate the individual component spectra completely by subtracting away the contributions from the cool-giant spectrum, as described by Griffin (1986). In Fig. 1.2 the same series of spectra as those illustrated in Fig. 1.1 have had the spectrum of the primary star removed, leaving as the residue the cleaned spectrum of the hot star with phase-dependent chromospheric features superimposed. The revelations from even this simple process can be quite stunning.

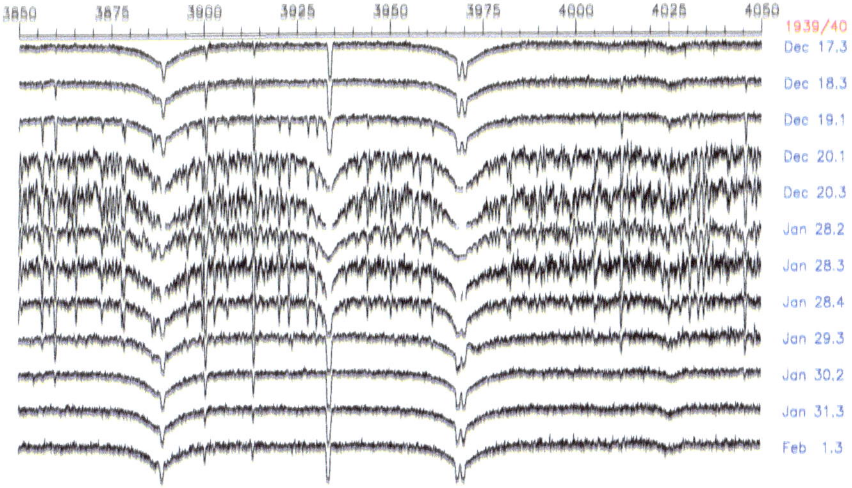

Fig. 1.2 Chromospheric features can be recognized and measured much more precisely when the blending lines in the primary's spectrum have been removed. This series of spectra is the same as that in Fig. 1.1, but the spectrum of the cool supergiant has now been removed, leaving the pure absorption signatures of the chromosphere on the almost smooth continuum of the mid-B secondary star. The original plates were digitized at the DAO

as Fig. 2.21 (page 49) shows. The development and honing of this procedure constituted a radical advantage for interpreting chromospheric spectra.

How well the subtraction technique works in a given case depends upon how closely the spectrum of the cool-giant component can be matched by a standard; if a match is not fully adequate in spectral type (temperature), luminosity class (log g) or—especially—velocity rest-frame, the output will be contaminated by 'mis-match' features. However, in the case of *eclipsing* composite-spectrum binaries the spectrum of the system during total eclipse is that of the cool star alone, thus providing *the* correct template for subtraction.

Nature also imposes a few troublesome limitations on this light-probe technique. While the classical systems ζ Aur, 31 Cyg and 32 Cyg can be investigated very effectively by this method, the luminosity ratio of the component stars in VV Cep and 22 Vul are so extreme that, in the spectral regions accessible from the ground, the signatures of both the hot companion and the chromosphere are overwhelmed by the flux from the primary star. At the other end of the scale, the chromosphere of a G-type giant is much smaller in extent in relation to the projected diameter of the hot secondary. In those cases the degree of spatial resolution of the chromosphere that can be achieved is greatly reduced, and the time resolution may be limited to a single night of chromospheric eclipse, or to the few hours when the secondary is in partial eclipse. That has tended to be the case with all the systems with normal and even bright G-giant primaries.

1.8 Stellar Chromospheres: Updating Information

Despite the progress in the study of ζ Aur binaries that has been achieved through the advent of computers and the acquisition of spectra at space-UV wavelengths, as well as CCD observations in the visible region with high signal-to-noise ratios, Wright's *Vistas* article has remained the only detailed reference work that is dedicated to the ζ Aur systems. A conference held by the American Astronomical Society in Monterey, California, focused almost exclusively on the ζ Aur stars. It was the first to do so for decades, and delegates felt it was high time to update Wright's comprehensive article with descriptions and discussions of the other systems that had since been discovered, and also with the new spectra of the classical systems that had been observed across a much broader span of the electromagnetic spectrum, the new observing technologies, the new approaches to interpreting the many features observed in these systems, the new theoretical models for chromospheric structure and support, wind and mass loss, and the new ideas for applying the knowledge thus derived to other stellar objects. The intention was primarily to update, rather than to provide a set of solutions to the many problems stated in Wright's article. Indeed, solutions to many of those problems are still beyond reach; it seems that the more carefully and closely we observe these systems, the more baffling are the details that we uncover and the more sophisticated the modelling that their interpretations demand.

This book is what has emerged from that decision. But though triggered by a conference, it is not the equivalent of a conference Proceedings, where each presenter reproduces a version of what s/he said, or intended to say, without necessarily much reference to other presenters and without including any of the (often so valuable!) discussions about what had been presented. Rather, what we have sought to produce is a unified collaboration between various of the specialists who attended the meeting. In reviewing progress from the vantage point of new technologies, new techniques and new ideas, and taking advantage of revolutionary achievements concerning data, tools and theory, the book acknowledges that to get the right answers one must ask the right questions: "Does binarity affect the structure of the chromosphere? Does it affect mass outflow? Can structure in the chromosphere be detected, and—if so—at what heights? How is the chromosphere heated? Are the energy requirements different for single stars?" Each of those questions would fill a conference on its own, and it is the endeavours to get at the truth which continue to drive research into such well-observed binary systems as these. In that regard, the book attempts to be a stepping-stone in the right direction rather than just a milestone marker.

Be that as it may, it bears the same title as the Monterey conference: *Giants of Eclipse*. The reason for the somewhat cryptic wording is our inclusion of the baffling system ϵ Aur. ϵ Aur shares hardly any of the physical properties of the ζ Aur systems, though it does manifest eclipses. Brighter, and therefore easier to observe, than any of the ζ Aur systems but with a period of 27 years, ϵ Aur has only had two, possibly three, eclipses in most of our working-life memories, and its actual properties remain largely a mystery. Since the mysterious opaque disk of ϵ Aur does not have a detectable spectrum of its own, information about its temperature, density and sub-structure has to be prised out of the signatures which it adds to the composite spectrum of the system at eclipse phases, particularly during ingress and egress. In that respect ϵ Aur could be said to resemble a ζ Aur binary, yet while it has often been associated with that group in the past, it has mostly been on account of the many puzzling *differences* rather than through any apparent, but misconceived, physical parallels. Programmes to monitor ϵ Aur have therefore concentrated on the epochs of eclipse (another close parallel with the ζ Aur stars!), though—as also with the ζ Aur systems—long-term monitoring throughout a whole orbital cycle is also needed.

The fact that the most recent 2-year-long eclipse of ϵ Aur, that of 2009–2011, was still very fresh in the memory, and many data had yet to be aired in public, provided ample incentive to include it in the Conference. It was also agreed that such a 'giant of eclipse' rightly deserved a chapter to itself in this book, which we have dedicated posthumously to K.O. Wright, leader and pioneer in the field and himself a giant of eclipse spectroscopy.

References

Ake, T.B., Fekel, F.C., Hall, D.S., et al: IBVS **2847** (1986)
Ake, T.B., Parsons, S.B.: IBVS **3002** (1987)
Beer, A.: MNRAS **95**, 24 (1934)
Bennett, P.D., Harper, G.M., Brown, A., et al.: ApJ **471**, 454 (1996)
Bottlinger, K.F.: AN **226**, 239 (1926)
Brown, A., Bennett, P.D., Baade, R., et al: AJ **122**, 392 (2001)
Burnashev, V.O., Burnasheva, B.A.: Bull. CRAO **107**, 1 (2011)
Demircan, O., Derman, E., Akalin, A.: IBVS **3486** (1990)
Eaton, J.A., Bell, C.: AJ **108**, 2276 (1994)
Faraggiana, R.: ZfA **62**, 99 (1965)
Gaposchkin, S.: HCO Circ. **421**, 1 (1937)
Griffin, R.E.M.: JRASC **82**, 49 (1988)
Griffin. R.E.M., Observatory **111**, 248 (1991a)
Griffin, R.E.M., Schröder, K.-P., Misch, A., et al.: A&A **254**, 289 (1992)
Griffin, R.E.M., Hünsch, M., Marshall, K.P., et al.: A&A **274**, 225 (1993)
Griffin, R.E.M., Marshall, K.P., Griffin, R.F., et al.: A&A **301**, 217 (1995)
Griffin, R.F.: S&T **81** 598 (1991b)
Griffin, R. & R.: JA&A **7**, 195 (1986)
Griffin, R. & R., Lines, H.C., et al.: IAPPP Comm. **44**, 82 (1991)
Guthnick, P.: AN **252**, 13 (1934)
Harper, W.E.: Publ. DAO **3**, 151 (1924)
Hynek, J.A.: ApJ **95**, 324 (1942)
Kron, G.E.: PASP **52**, 124 (1940)
Lee, E.K., Wright, K.O.: Publ. DAO **11**, 339 (1960)
McKellar, A., Petrie, R.M.: MNRAS **112**, 641 (1952)
McLaughlin, D.B.: ApJ **111**, 449 (1950a)
McLaughlin, D.B.: PASP **62**, 13 (1950b)
Menzel, D.H.: HCO Circ. **417**, 1 (1936)
Parsons, S.B., Ake, T.B.: IBVS **2334** (1983)
Parsons, S.B., Ake, T.B.: BAAS **16R**, 913 (1984)
Richardson, E.H.: JRASC **62**, 31 (1968)
Saitō, M., Sato, H., Saijo, K., et al.: PASJ **32**, 163 (1980)
Schröder, K.P., Pols, O.R., Eggleton, P.P.: MNRAS **285**, 696 (1997)
Walker, G.A.H., Walker, A.R., Irwin, A.W., et al.: BAAS **27**, 1430 (1995)
Wright, K.O.: VA **12**, 147 (1970)
Wright, K.O., Lee, E.K.: PASP **68**, 17 (1956)
Wright, K.O., Odgers, G.J.: JRASC **56**, 149 (1962)
Wright, K.O., Hesse, K.E.: Publ. DAO **13**, 301 (1968)

Chapter 2
Observing and Analyzing the ζ Aurigae Systems

R. Elizabeth Griffin, Joel A. Eaton, Thomas B. Ake, and Klaus-Peter Schröder*

Abstract This chapter reviews the scientific information that has been extracted from the ζ Aur binaries by various methods, surveys the kinds of spectroscopic data that have been used in those analyses, and examines each of the 10 known eclipsing systems in turn, except for VV Cep which is treated separately in Chap. 3. It describes what has been achieved by modelling, and assesses what has been learned so far about the mass loss and winds from the cool giants and supergiants in those systems. It tries to keep in focus the fundamental questions posed at the end of the previous chapter (Sect. 1.8), in particular the role of the ζ Aur binaries as templates for single stars, and suggests observing strategies for the future.

2.1 Preamble

The study of stellar chromospheres is first and foremost phenomenological. The solar chromosphere would not have been so easy to observe had it not been for the truly remarkable coincidence that the projected disk of the moon is very nearly the

*Contributions have also been made by Phil Bennett, Roger Griffin, Wendy Hagen Bauer and Graham Harper

R.E. Griffin (✉)
Dominion Astrophysical Observatory, National Research Council of Canada, 5071 West Saanich Road, Victoria, BC V9E 2E7, Canada
e-mail: Elizabeth.Griffin@nrc-cnrc.gc.ca

J.A. Eaton
7050 Bakerville Rd., Waverly, TN 37185, USA
e-mail: eatonjoel@yahoo.com

T.B. Ake
Space Telescope Science Institute/CSC, 3700 San Martin Drive, Baltimore, MD 21218, USA
e-mail: take@stsci.edu

K.-P. Schröder
Department of Astronomy, University of Guanajuato, Apartado Postal 144, 36000 Guanajuato, Gto, Mexico
e-mail: kps@astro.ugto.mx

T.B. Ake, E. Griffin (eds.), *Giants of Eclipse: The ζ Aurigae Stars and Other Binary Systems*, Astrophysics and Space Science Library 408,
DOI 10.1007/978-3-319-09198-3_2

same size as that of the Sun. Although the presence of a chromosphere around a star could have been *presumed* by analogy with the Sun, all that changed with the sight of those first atmospheric-eclipse spectra of ζ Aur in 1932. Photometry and spectroscopy of ζ Aur, and of other similar eclipsing binaries, have been carried out fairly frequently ever since, with concentrations near eclipse phases, and hopes of discovering more members of the class have not been allowed to die. Fascinating spectra at shorter-wavelength UV regions were obtained from space with *IUE*, in operation from 1978 to 1996 and supplying astrophysics with many surprises as well as extremely valuable data. More spectra from space, of higher quality but fewer in number and more limited in time-span, were procured with the *Hubble Space Telescope*, while forays into the shorter-wavelength UV by *FUSE* clinched the evidence that the component stars in even these nominally detached systems interact at some level, plus unique information concerning the energies and processes that are involved.

Recognizing that a binary eclipses, and then refining what is known about the system and its component stars, is achieved through a synergy of different forms of observations. Ideally those include (a) photometry, for accurate eclipse timings, and estimates of the radii of the component stars, (b) spectroscopy and spectrophotometry, for the luminosities and other physical properties of the individual stars, and for specific details of the chromosphere, (c) radial velocities, to determine the relative velocities of the two stars and for modelling the geometry of the eclipse, (d) interferometry, to determine orbital inclination; a high value signals the likelihood of eclipses, as was true for both τ Per and γ Per, (e) imaging, if possible (actually *seeing* what was causing the eclipse in ϵ Aur—though not a ζ Aur binary—settled one aspect of a decades-old argument), and (f) asteroseismology, to probe the evolutionary history of the giant star and assess its age and mass. Since the partial-eclipse phases of supergiant systems may last for more than a single night, full coverage of an eclipse light curve requires a network of photometric observers at different longitudes.

Each atmospheric-eclipse spectrum corresponds to a particular point along the projected trajectory of the secondary, and therefore to a specific height in the primary's chromosphere determined from a model of the eclipse geometry. By measuring and combining the strengths of the chromospheric absorption lines one can derive the temperature and density gradients in the chromosphere, at least for the material in the neighbourhood of the secondary's projected trajectory.

The classical ζ Aur systems (31 Cyg, 32 Cyg, ζ Aur itself and VV Cep), which have been known and studied for over 60 years, have K or M supergiant primaries and relatively long orbital periods. In contrast, most of the subsequent additions to the class have G-type giant or bright-giant primaries, and present a rather different set of challenges and limitations. In that respect the ζ Aur systems fall into two groups: those with K-supergiant primaries that exhibit massive chromospheres extending to heights comparable to the supergiant's radius and column densities that reach high values, and those with G-type primaries with geometrically much thinner chromospheres (reaching only to heights of some 0.1 of the giant's radius, or less) and much smaller column densities. When plotted on an H–R diagram, all the primary stars—except for 22 Vul—cluster into two distinct groups on either side of the 'dividing line' of Linsky and Haisch (1979); 22 Vul is unique, in that it seems to be a borderline case. Sections 2.6.1 and 2.6.2 compare and contrast the chromospheric characteristics of the various ζ Aur primaries in relation to that 'dividing line', but there is not yet a convincing physical explanation beyond it being a convenient phenomenological descriptor.

The division just described has consequences both for what we can learn from atmospheric eclipses, and for how we must observe them. In the classical ζ Aur systems the orbital parameters and relative sizes of the component stars are such that the projected distance through which the hot secondary moves in a day is of the order of its own diameter, maybe $\sim$5% of the radius of the supergiant. Since the density gradient in the inner regions of the supergiant's chromosphere is generally quite steep, the difference between line-of-sight columns on adjacent nights is easily discernible. In contrast, in the systems with G-type giant primaries the component stars are less disparate in size, the giant's chromosphere is less extended and the giant is somewhat hotter, with the nett result that the strong absorption features which characterize the atmospheric eclipses of a classical system are much less pronounced. In this G-type group, not only is an eclipse much more rapid, but the spatial resolution is inevitably much coarser. If the projected disk of the secondary is larger than the physical extent of the chromosphere, then when the secondary is fully out of geometrical eclipse only part of our line of sight to it traverses the giant's chromosphere, so the chromospheric absorption spectrum that is observed is somewhat diluted. That lack of spatial resolution may compromise the scientific usefulness of the observation without an adequate model of the density gradient. Conversely, during partial eclipse the line of sight to the secondary will sample just the chromosphere, but the flux from the secondary is then sufficiently reduced that the signal-to-noise (S/N) of the observation is seriously reduced as well. Nevertheless, the broader range of cool-star types and chromospheric properties which can now be studied through this technique of 'light probing' (sometimes referred to as 'eclipse mapping') makes the study of *all* of these eclipsing systems very attractive indeed.

Observations of the Ca II *H* and *K* region have been, and still are, central to ground-based spectroscopy of ζ Aur systems. But while the many published studies by Wilson (such as Wilson and Abt 1954) and by Wright (culminating in Wright 1970) remain magisterial products of a classical age, Wilson expressed frustration at the limitations which were inherent in the spectroscopy of those decades when he concluded: "In the light of what has been done thus far, I am unable to see how extensive further observation of eclipses of ζ Aur can add much to our basic knowledge of how its chromosphere is constructed. The same absorption lines will simply be observed over again and unless some new approach in the method of analysis is uncovered, it is difficult to see that much progress could result." (Wilson 1957). In fact, much ground-based monitoring continues to concentrate on the *K* line because continuity is an essential aspect of long-term studies. Furthermore, as this chapter goes on to show, computers and space observations have rejuvenated the field to an impressive degree.

2.2 Spectroscopic Observations: Variety and Complementarity

2.2.1 Ground-Based Spectra: Historic ('Heritage') Data

When K.O. Wright published his survey and review of the ζ Aur systems (Wright 1970), astronomy's workhorse detector was the photographic plate—excellent in many respects, even though its non-digital nature seriously curtailed the operations and analyses which were later to become possible with born-digital spectra. Seven of the discoveries of the 10 known bright ζ Aur systems were in fact made with ground-based spectrographs working in photographic mode. By the 1960–1970s technology to convert photographic observations into digital spectra was commencing, but digital computing was also only in its infancy and even basic manipulations such as co-adding spectra to increase the S/N were slow and expensive. Spectroscopy from space was still in the future, so information about the astrophysics of the ζ Aur systems was restricted to what could be observed from the ground.

The acute interest in ζ Aur systems that prevailed at Mount Wilson (California) and at the DAO (Canada) ensured that the plate archives at those observatories, at least, are well stocked with high-quality, high-dispersion spectra of the first four ζ Aur systems to be discovered and studied. Smaller-format spectra of ζ Aur binaries were also obtained and analyzed by active groups in (inter alia) Michigan (USA), Stockholm (Sweden), Merate (Italy), Hamburg and Bonn (Germany); in the absence of any digitized inventories the exact facts and figures are difficult to come by. Photometric measurements were—and still are—carried out at many sites worldwide, mostly by individuals and often by amateurs; published announcements date back more than 80 years. Between 1987 and 1995 high-dispersion spectra of several ζ Aur binaries were obtained photographically at Calar Alto Observatory (Spain).

One enviable property of the photographic plate is its physical longevity. When chemically fixed, the image remains in its virgin state, as readable now as it was then, and only physical damage can destroy it. (Who can say the same of electronic images?) The main disadvantage, in the context of observational astrophysics, is its intrinsic non-digital nature. Today's researchers require their data to be accessible electronically, so analogue data are therefore effectively invisible as sources of information, however uniquely they may in fact complement more modern ones. That technical disconnect between pre-digital and born-digital data has been allowed to become a physical boundary to the time-domain of scientific research, much to the detriment of (in particular) studies of objects that undergo long-term changes that span years or decades. Most working archives of ground-based electronic data do not go back more than about 20 years, and research into objects like the ζ Aur binaries is consequently impoverished in certain aspects.

Fortunately, that technology divide can now be bridged with state-of-the-art equipment, and once a photographic observation has been transformed correctly into an electronic one it can be used just as if it had been acquired digitally in the first place. Digitized photographic data offer 'instant history' that has irrefutable value when timescales are long; both VV Cep ($P \approx 20$ years) and 31 Cyg ($P =$ 10.4 years) fall into that category, as also does the 27-year system ϵ Aur (Chap. 4). In the case of ϵ Aur, digitizing heritage spectra was the *only* way to expand the time-base of electronic observations in order to analyze and intercompare different eclipses (Griffin and Stencel 2013). Procedures for digitizing photographic spectra correctly are outlined in the Appendix (page 80).

2.2.2 *The Switch to Digital Technology*

With the switch to digital observing, electronic data management and computer-based modelling, astronomy underwent a revolution whose ramifications were at least as shattering as those experienced when photography superseded visual observing in the late nineteenth century. The nature of the switch was sufficiently abrupt, distinctive and universal that it has come to be regarded retrospectively as the commencement of a new scientific era, but it has had the more unfortunate consequence that photographic data have become relegated to 'history' and frequently dismissed from the context of modern research. Yet the date of the switch is quite artificial. Pre-electronic data have an on-going relevance to time-domain research that is unquestionable, particularly in the context of decades-long variability.

The changeover from photography to digital detectors such as the CCD did not proceed seamlessly. For an interesting period of 10 or more years the use of the CCD in astronomy outpaced the equally necessary development of matching computer speed and capacity. Files of spectra either accumulated on magnetic tapes or occupied volumes of storage that were significantly larger than most machines or archives could accommodate in any number, and were discouragingly expensive as well. Data storage and archiving as we know it today was simply not possible, and was not even entertained as a necessary adjunct to observing. The nett result is a rather sorry gap of several years in astronomical archives worldwide, and helps to explain an apparent lack of archived spectra of ζ Aur systems during that period of transition. Fortunately, attitudes relaxed as computer power increased and information technology advanced.

A few key spectroscopic archives, such as those of ELODIE and SOPHIE at Haute-Provence (France), now offer stellar spectra in fully reduced format, but most others still have a long way to go before their contents are as immediately accessible as the vision of astronomy's 'Virtual Observatory' requires. Matters are more organized at national observatories, but retrieving fully calibrated ground-based spectra from observatory archives is still mostly a thing of the future. Accessing such spectra of ζ Aur systems is therefore not something that can be accomplished readily, if at all. All the optical spectra discussed in this chapter were in fact observed by the authors or collaborators.

The studies reported by Wright in 1970 were inevitably limited by close blending between the lines of the cool-giant primary and similar ones formed in the chromosphere. Because the inner layers of the chromosphere are effectively an extension of the photosphere and at a similar temperature, many of the lines that appear during a chromospheric eclipse are the same low-excitation and ionization lines as are dominant in the primary's spectrum. Having spectra in digital format opened a host of opportunities for separating those blends, measuring line-strengths and line-positions, or matching observed spectra to synthetic ones derived from model-atmosphere calculations. In the case of composite-spectrum binaries, the ability to subtract away the spectrum of the cool-giant primary (usually the dominant component) offers a powerful tool for isolating the spectrum of the hot star and enabling its radial velocity (RV) to be measured as accurately as the nature of its spectrum permits, and yielding the mass ratio of the system quite precisely. The subtraction tool, discussed in detail by Griffin (1986), has been applied to individual ζ Aur binaries by Griffin et al. (1990, 1992, 1993).

2.2.3 *Observations from Space*

Observations of high-resolution UV spectra from space, first made possible in the 1970s, opened a whole new chapter in the study of ζ Aur systems. The flux from the cool supergiant primary is nearly (though not quite) negligible shortward of about λ 2800 Å, so UV spectra are dominated by the flux from the hot component. The UV region contains many resonance lines that originate in the chromospheric gas, and some that come from low-lying metastable states. Interactions between the hot secondary and its cool primary are also much more active at higher energies, so in the UV there are altogether many more processes to diagnose and spectroscopic features to examine than at optical wavelengths. Against that, the continuum photons from the hot star are scattered by atomic transitions in the circumstellar gas around the cool primary star, giving rise to an additional emission-line spectrum that complicates each total-eclipse observation and—for some stars—the observations at all other phases too. Strong interstellar lines of Mg II and other species, possibly as multiple components, may also be present, and add to the confusion. The space-UV region can therefore yield rich information about a stellar chromosphere and the immediate environment of the giant and of the binary system, though identifying and unravelling the individual contributions may sometimes be challenging.

Space-UV spectroscopy commenced with *OAO-3* (also named *Copernicus*), which operated from 1972 to 1981 at resolutions of up to 0.05 Å. Though limited to fairly bright stars, *Copernicus* revealed for the first time some of the wealth to be found in UV stellar spectra. Its successor, *IUE*, did not exceed the spectral resolution of Copernicus but could reach a far greater sample of stars. Launched in

1978 and lasting for nearly 19 years, *IUE* carried two spectrographs (for λ 1150–2000 Å and 1850–3300 Å), each offering a high- and a low-resolution mode of 0.2 and 6 Å resolution, respectively. That spectral region proved a bonanza for studies of the ζ Aur binaries; it included the *h* and *k* lines of Mg II near λ 2800 Å, of C IV near λ 1550 Å and Si IV near λ 1400 Å (both of which contain specific information about the ionizing winds of the hot companion of a ζ Aur binary), though many of the crowds of other atomic lines which it also featured could not then be classified. Several 'epochs' of observing time on space telescopes were awarded to programmes on ζ Aur binaries, at first mostly to observe eclipse phases but later to investigate other phases, which early results had showed to be necessary too.

Space-based observations offered many specific advantageous features that could complement ground-based spectroscopy:

- Strong lines of ionized metals, often up to the third ionization stage, furnishing information on the constituents of the chromosphere and on electron densities,
- Hydrogen lines, observed directly as Lyα absorption,
- Many Fe II multiplets, for deriving excitation temperatures,
- Wind absorptions, yielding kinematic information about wind flows at great heights,
- A rich spectrum of scattering, informing about the wind kinematics, and
- Lines that are intrinsically too weak to be scattered in the wind, and which yield the thermal and density structure of the chromosphere.

As described by Parsons and Ake (1998), UV spectroscopy also provided a wealth of data for single-lined (SB1) binaries that exhibit composite spectra, indicating that a cool evolved star was paired with a hot but much fainter one—too faint to be detected at all easily at visible wavelengths in the presence of the giant's dominating spectrum. Determining the mass ratio of such systems, which depended on measuring both spectra, was therefore best carried out in the UV where the fluxes from the two component stars were more nearly equal.

IUE was the first telescope to offer a public archive of *fully reduced* spectra, a feature which proved so popular that, on average, each *IUE* observation was downloaded five times by the astronomy community. The *IUE* Final Archive was an innovation which paved the way not only for space-mission archives but for ground-based ones too; 'saving the bits' is an essential first step, but someone not intimately familiar with the equipment is unlikely to treat the raw files as correctly as would a pipeline that had been prepared by the team that designed and built the spectrograph—or even find it profitable to try.

During its operating lifetime *IUE* observed all of the 10 known ζ Aur systems at eclipse phases except for γ Per, whose eclipses were not confirmed until one was observed in 1990 (the period is 14.6 years). Accumulations of data for the four coolest systems supported numerous efforts to model their chromospheres and winds. The analyses of ζ Aur (Chapman 1983; Eaton 1993a), 31 Cyg (Stencel et al. 1984; Eaton and Bell 1994), 32 Cyg (Stencel et al. 1979; Eaton 1993b) and of VV Cep (see Chap. 3) were all based on UV spectroscopy.

Spectra from GHRS, the Goddard High-Resolution Spectrograph on *HST* (the *Hubble Space Telescope*) from 1990 to 1997, sharpened and extended the results derived from *IUE* spectra by virtue of superior S/N and higher resolution. Operating at its maximum resolving power of 85,000, GHRS was fitted with stable photon-counting UV detectors rather than the vidicon detectors of *IUE*. The data it yielded were a new generation in terms of quality, spectral purity and calibration, making it possible to obtain high S/N observations of the chromospheric and wind-line profiles which *IUE* had only partially resolved. High-ranking among the many space-based programmes that are pertinent to this review was an extensive study of the 1993 eclipse of ζ Aur (Bennett et al. 1996; Baade et al. 1996).

The Space Telescope Imaging Spectrograph (STIS) followed GHRS for another 7 years until a power-supply failure in 2004, but was repaired and returned to service in 2009. STIS UV spectra extend from λ 1150 to 3150 Å in echelle mode, with a resolving power up to 114,000. The *HST* data archives contain raw data files from STIS, but calibrations are carried out 'on the fly' so files that are downloaded incorporate the most recent version of the instrument's calibration.

FUSE, the *Far Ultraviolet Spectroscopic Explorer*, was launched in 1999 and operated for 9 years, recording spectra in the far UV from λ 905 to 1187 Å (so just below the Lyman limit) with a resolving power of 20,000–30,000. *FUSE* offered a complementary approach to understanding ζ Aur stars through its ability to probe domains of hotter material in a supergiant atmosphere, particularly the upper transition region and lower corona.

Some ζ Aur binaries have also been detected at radio wavelengths through free-free emissions (thermal bremstrahlung) from their winds (Drake and Linsky 1986; Drake et al. 1987; Harper et al. 2005). The observations supported the high mass-loss rates of $10^{-8}\ M_\odot\ \text{year}^{-1}$ deduced from spectroscopy, and confirmed the slow acceleration of the winds which was inferred from modelling the space-UV line profiles and the scattering of light by the wind. In addition, non-spectroscopic tools such as interferometry (Chap. 6, page 157) and asteroseismology (Chap. 7, page 169) offer exciting and quite different means of extending our understanding of the interiors as well as the outer atmospheres of these stars.

Those new opportunities—the ability to observe from space, to isolate chromospheric signatures purely, to investigate phenomena by analyzing high-energy processes, to obtain high-quality data, to work robotically and to exploit the flexibility of electronic data formats—injected a new dimension into the observation, analysis, study and (literally) the modelling of the ζ Aur systems that carried forward and expanded the momentum initiated by the early researchers. Some early by-products were the discoveries of more eclipsing systems, and the recognition that others known to eclipse belong to the ζ Aur category. Even so, the sample of definite members is still far too small for an 'average' to have any scientific meaning—or is every star in fact the individual that it seems at present to be? Only more discoveries and more observations will demonstrate whether there is such an entity as 'a' stellar chromosphere, so in that respect this book can only be an interim report.

2.3 Analysis of a Chromosphere: Extracting Column Densities

Measured equivalent widths of chromospheric features of a given atom or ion yield the column density—the cumulative concentration (number density) of that element along a chromospheric column of width equal to the projected diameter of the hot star at the time of the observation. Connecting the different results through an adopted description of the temperature in the chromosphere leads to total column densities for each phase of the observed chromospheric eclipse. A set of column densities can then be combined, and tested against a density model. A starting model might follow an exponential drop with height, assume radial symmetry, and ignore local inhomogeneities. Experience from ground-based observations showed that, where the low chromosphere is sufficiently dense, an additional slowly-varying component may also be required in order to model the corresponding portions of the *B* and *U* photometric light curves.

The derivation of mass column densities and the physical properties of the gas that absorbs along a line of sight throughout the extended atmosphere is based on simple attenuation theory. At chromospheric heights where densities are sufficiently low, the expression of radiative transfer in a line becomes one of pure scattering, given ample radiative transition probability (gf values not much lower than 0.1). According to the unique geometry of a chromospheric eclipse, the companion star shines from far behind the chromosphere and subtends a small solid angle at the point of scattering, so very little of its light gets scattered towards the observer in comparison to the amount of absorption that there is in the same line.[1] That natural advantage was recognized and exploited by the pioneering researchers, mainly Wellmann (1939), Wilson and Abt (1954), and Groth (1957), who used a classical curve-of-growth approach for pure line absorption as described by Unsöld (1968). In that way it became possible to derive empirically the chromospheric column density of any given ion and excitation level which was represented by chromospheric absorption. Even though stellar chromospheres embrace a wide variety of parameters and situations, this attenuation theory is still adequate for determining a basic density model.

Section 2.1 identified two groups of binaries: those with K-supergiant primaries and those with G-type giant or bright-giant primaries. In the first group the chromospheric scale-height is much larger than the projected disk of the companion star, at least above the lowest layers, so to a first order the hot star may be approximated by a point light-source. Their more massive chromospheres produce very rich absorption-line spectra, for which the curve-of-growth approach should work well provided strongly saturated lines are avoided. Modelling the K-line on

[1]Somewhat fortunately, the low densities in these extended atmospheres—some 10^8 times less than in Earth's atmosphere—obviate the need to consider the refraction along the line of sight that complicates the transfer of starlight through planetary atmospheres during occultations, even with the assumption that the chromosphere is composed entirely of neutral hydrogen.

its own could in principle extract the same information as does a curve of growth based on a whole multiplet of lines, but when a line core is saturated it is difficult to determine the best-fitting parameters unambiguously.

On the other hand, the thin chromospheres of the second group can *only* be traced quantitatively in the optical region through the intrinsically strong Ca II K-line absorption. A further complication arises here because the chromospheric scale height is comparable to, or even smaller than, the projected diameter of the hot star. Modelling a line profile must then include a density gradient across the companion's projected disk (see Schröder and Hünsch 1992; Griffin et al. 1992; Schröder et al. 1994, 1996). In extreme cases, where the part of the projected disk nearest to the photosphere is viewed through strongly saturated line-absorption with damping wings while the outer parts of the same disk see an optically thin line, the resulting line profile may well be composed of shallow damping wings and an unsaturated core. Determining the most representative density gradient is clearly critical, though choices may be hampered by insufficient observations if an eclipse is very rapid.

Since neither electronic data nor electronic analytical tools were available to the early researchers, their analyses were based on the curve-of-growth method out of sheer necessity, and since the physical properties of the chromosphere change with height, those preliminary models can only be first-order approximations. With modern computing tools we can now calculate, for any specified temperature gradient, the density, turbulence, expansion velocity, ionization or departure from LTE, and the attenuation of light along a line of sight. Fine details can be tested if the observed line profiles are well resolved, as they are in *HST* spectra (e.g., Eaton 1988, 1993a,b; see also Chap. 3).

2.4 Individual ζ Aur Systems

A major emphasis of Wright's (1970) review of the ζ Aur binaries was the physical properties of the systems and their component stars. We now have a better idea of many of those parameters and their true uncertainties, and the following pages discuss relevant features about the individual systems in turn. In some cases, improvements can be achieved by including the ellipsoidal variability of the cool-giant components (e.g., Eaton and Shaw 2007; Eaton et al. 2008); in other cases, the availability of more precise radial velocities is enabling a significant refinement of the orbital elements (e.g., Griffin 2005, 2008). Difficulties in deriving precise measurements for the hot secondary components, particularly when that component is somewhat earlier than A0, inevitably beget substantial uncertainties in measurements of the system's mass ratio, so the masses of the components of the classical systems are consequently much more poorly known than one would like. However, the situation is encouragingly good for the more recently discovered ones, notably HR 6902, γ Per and HD 223971 (Sects. 2.4.6, 2.4.9, and 2.4.10, respectively). Those binaries are obviously double-lined, their secondary stars are not rotating fast, and their spectra show lines of metals as well as of hydrogen.

This section addresses each of the ζ Aur systems in turn, offering a brief history of how discoveries of chromospheric eclipses came about, a review of some of the more recent results, and samples of chromospheric absorption spectra to illustrate the immense variety which they present as a group. An exception is made of VV Cep, which is sufficiently complex, and has such a wealth of UV data to discuss, that it is treated separately in Chap. 3 (page 85). It also gives a very brief introduction to ϵ Aur, which is treated fully in Chap. 4 (page 107). The basic parameters of the ten ζ Aur systems are set out in Table 1.1 (page 7).

2.4.1 ζ Aur

ζ Aur, the eponymous member of the class, has been studied extensively since the discovery of its atmospheric eclipse in 1932 (Guthnick 1934). The first paper in the Astrophysical Journal Supplement Series presented a detailed analysis of ground-based spectra of the atmospheric eclipse of this binary (Wilson and Abt 1954), and it has gone on to be the subject of extensive studies based on space data (e.g., Bennett et al. 1995, 1996; Baade et al. 1996). Papers by Griffin et al. (1990) and Griffin (2005) summarize adequately the history of observing and analyzing this star. The duration of totality is about 36 days. Figure 2.1 offers a scaled geometrical model of the system; it adopted a supergiant's radius, R_K, of 154 $R_\odot$ (Di Benedetto and Ferluga 1990).

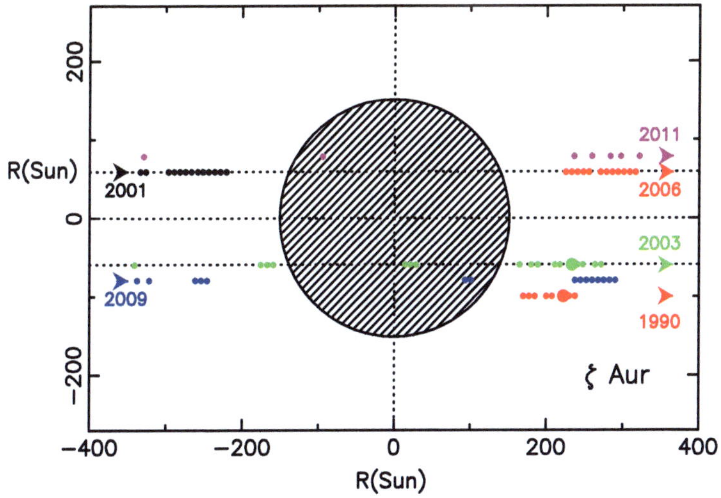

Fig. 2.1 Scaled geometry of ζ Aur, indicating the monitoring of six eclipses. The *large dots* refer to the onset of line-splitting (see text). The modelled trajectories have been displaced vertically for clarity, and should all lie along the same one. 1990 data: photographic spectra from Calar Alto Observatory; others: CCD spectra from the DAO. *Observations by Griffin*

The orbit of the K supergiant is relatively well known, and recent solutions tend to improve the precision of the elements rather than offer fresh values. Griffin (2005) gives the period as 972.164 ± 0.041 days, or very close to $2\frac{2}{3}$ years. However, an accurate double-lined solution is more elusive because the spectrum of the mid-B secondary contains few lines (besides hydrogen lines) that can be used for RV measurement. Moreover, all are broadened by rotation and are contaminated by chromospheric absorption, by the overlapping spectrum of the K star, and (in the UV) by emission that is scattered in the expanding atmosphere and wind. Attempts to detect precise RV signals from the B-type component in ground-based and satellite-UV spectra have yielded values for the mass ratio, $q = M_1/M_2$, that are disconcertingly scattered, though Bennett et al. (1995), using UV lines in GHRS spectra, derived a surprisingly precise velocity amplitude for the B star that corresponded to $q = 1.21 \pm 0.02$, with individual masses of $M_K = 5.8 \pm 0.02\ M_\odot$, $M_B = 4.8 \pm 0.02\ M_\odot$. Attempts by Schröder et al. (1997) to test stellar-evolution theory by comparing the positions of the component stars of ζ Aur in the H–R diagram with theoretical evolutionary tracks encountered some ambiguity which seemed to hinge on the level of inherent uncertainty in the mass of the hot secondary.

2.4.1.1 Ground-Based Monitoring

In the optical spectrum the Ca II H and K lines provide a rich description of the chromosphere of ζ Aur. The changes that can be recorded by monitoring those lines during ingress and egress have been the traditional focus for ground-based spectroscopy; even when observations were photographic, and a spectrum in the blue or near-UV spanned several hundred Ångströms as well as the H and K region, investigations still tended to concentrate on the K line—as, too, do numerous recent CCD studies. The high relative abundance of Ca and the great intrinsic strength of the H and K features at the temperatures prevailing in a chromosphere mean that those lines can be detected at great heights above the photosphere when other features have grown too weak to be discerned. But while saturation in the lower chromosphere means that the H and K lines are too strong for useful measurement or analysis, there are also many intrinsically weak chromospheric features that can only be detected in the denser (lower) chromospheric layers. Those are sampled best during phases of deep chromospheric and partial eclipse, when the hot star shines through the boundary between the chromosphere and the photosphere (see Fig. 2.2).

Griffin et al. (1990) extracted spectra of the low chromosphere and measured the strengths of 260 identified lines, which Schröder et al. (1990) then analyzed by a curve-of-growth method, deriving hydrogen particle densities of $\sim 5 \times 10^{11}$ cm^{-3} at physical heights between 3 and 9 $R_\odot$. That result is pursued further in Sects. 2.6.1 and 2.9 in relation to the commonalities between individual supergiant chromospheres, and whether a ζ Aur star can be a valid template for single K supergiants. Though lending its name to the whole class of binaries that manifest atmospheric eclipses, ζ Aur is by no means representative of any kind of norm or average.

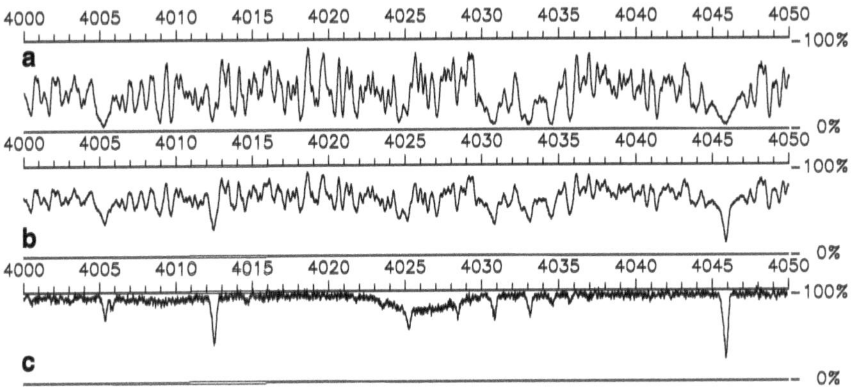

Fig. 2.2 Uncovering a chromospheric spectrum in ζ Aur. The spectrum of the supergiant alone (**a**), observed during the 1987 total eclipse, has been subtracted from the composite spectrum (**b**) observed during ingress, revealing (**c**) the secondary spectrum and numerous sharp chromospheric lines. The broad feature is He I at λ 4026.2 Å (Data source: Calar Alto Observatory photographic spectra, observed by Griffin (*Reproduced, with kind permission, from Griffin et al. (1990)*)

Considerably less turbulent that the chromospheres of either 31 Cyg or 32 Cyg (taking line-widths as the guide), its own chromosphere also exhibits far fewer of the 'satellite' features that have characterized the appearance of the *K* lines in 31 and 32 Cyg.

Nevertheless, ζ Aur also harbours some fascinating surprises, one of which was revealed in 1990. By chance, a 12-night observing run at Calar Alto Observatory in August–September 1990 to monitor the *K* line of 32 Cyg during late ingress coincided with egress phases of ζ Aur. Despite its unfavourable position in the sky then, it was possible not only to obtain a much more complete sequence than was usually possible during its winter eclipses, but also to continue sampling its chromosphere at substantially greater heights than had been attempted before. The result was startling, as Fig. 2.3a shows.

The curious splitting of the *K* line that commenced after 10 or so days beyond 4th contact in 1990 does not appear to have been seen before, probably for the simple reason that observers did not routinely pursue chromospheric effects in ζ Aur to such large distances (days) from first or fourth contacts. Comparable splitting and similar asymmetry was seen in the Ca II *H* line, and also in the low-excitation Ti II lines at λ 3759, 3761 and 3685 Å even though they are intrinsically much weaker than the Ca II lines. However, it is not a regular feature; it was seen again in 2003 (Fig. 2.4a) but not in either 1998 (Fig. 2.3b) or 2009 (Fig. 2.4b).

It had been inferred as long ago as 1954 by Wilson and Abt (1954) that a chromosphere could be 'clumpy', and several publications describe transient satellite components to the *K* line at randomly high velocities, but it is not at all obvious that clumps or clouds are what caused the line splitting illustrated here. 'Clouds' can be detected moving in just 1–3 days through the *K*-line profile in both 31 Cyg and 32 Cyg, and also in ζ Aur though much less frequently, but those seem to

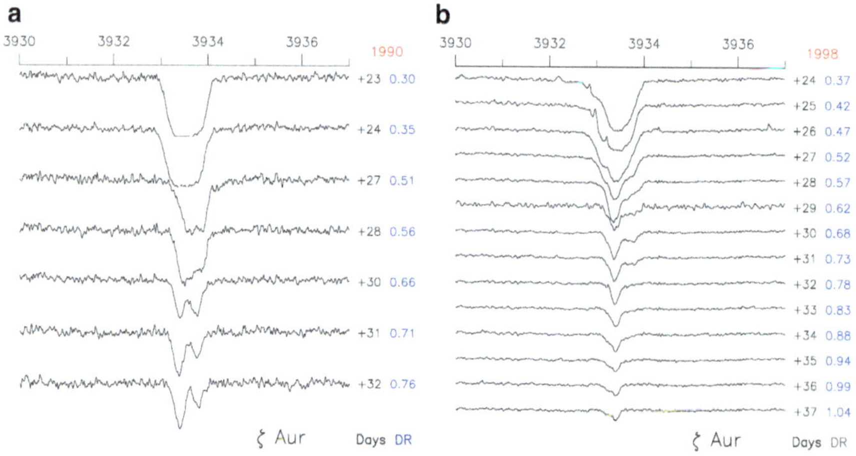

Fig. 2.3 The chromospheric K line in ζ Aur during egress in (**a**) 1990 and (**b**) 1998. The spectrum of the supergiant primary has been removed. 'Days' is the number of days since mid-eclipse; DR is the height from the photosphere in K-star radii, R_K (Data sources: (**a**), Calar Alto Observatory, photographic spectra; (**b**), DAO, CCD spectra. Observations by Griffin)

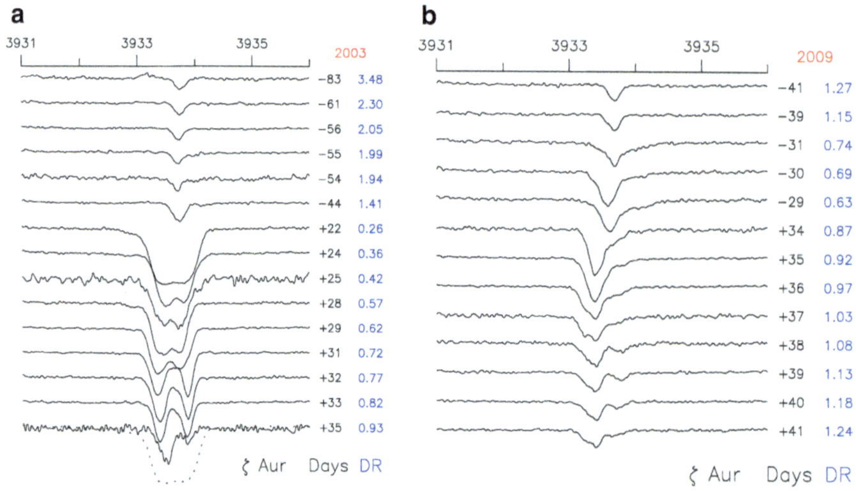

Fig. 2.4 The chromospheric K line in ζ Aur during ingress and egress in 2003 (**a**) and 2009 (**b**). The spectrum of the K supergiant has been removed. 'Days' and DR are as for Fig. 2.3. Both plots have the same vertical scale; on day +22 in 2003 the K-line core was 98% deep, and is repeated (*dotted*) at day +35 for reference (Data source: DAO CCD spectra, observed by Griffin)

be independent events whereas the profiles of the split lines revealed approximately the same velocity structure for a number of days. The earlier observers also invoked 'condensations' in the chromosphere to explain the mysteriously high electron densities which they deduced for the high chromosphere, but those seem

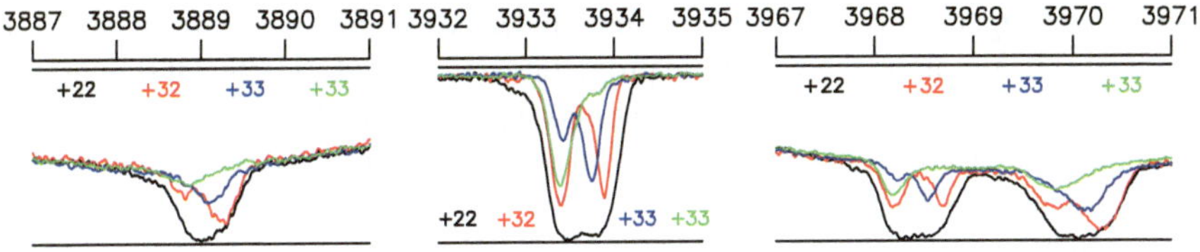

Fig. 2.5 Comparing chromospheric profiles of H_8 (*left*), Ca II K (*centre*) and Ca II $H+H_\epsilon$ (*right*) in ζ Aur during the same egress phase in 2003 (*red*), 2006 (*blue*) and 2009 (*green*). The *black profiles* correspond to an earlier egress phase in 2003. The numbers give the intervals, in days, since central eclipse (Data source: DAO CCD spectra, observed by Griffin)

to refer a different phenomenon that was ubiquitous and persistent throughout the chromosphere. The splitting seen in these optical spectra of ζ Aur does not answer to those requirements and seems to be of a different nature, since it is as often absent as present. Moreover, when splitting *does* occur, the overall chromospheric K-line profile is noticeably stronger than when it does *not* occur. Its occurrence therefore has an prominent intermittent element.

The observations represented in Fig. 2.1 were timed in order to determine the range of heights where line-splitting might occur. So far, in optical spectra it has been detected only during egress, and only between ~0.4 and 1.1 R_K above the K-star's limb. The split components show a velocity separation that is regular and well constrained, and they always keep within the profile of the saturated one formed lower down (as illustrated in Fig. 2.5), and by the dotted one observed on day +22 in Fig. 2.4a and repeated for comparison at day +35. Similar line-splitting has also been detected in ζ Aur in UV GHRS spectra at comparable phases, thus reinforcing the suggestion that the cause lies in the structure of the chromosphere. One tentative suggestion is that the splitting is caused by the superposition of emission, rather than by discrete absorption components that suddenly move with similar velocities but of opposite sign. Since the line of sight is sampling the chromosphere *tangentially*, such emission has mostly only radial motion. That interpretation is supported by the fact that the split components keep within the overall profile of the chromospheric K-line profile (as seen, for instance at a height of +22 days in Fig. 2.5), and is an interesting application of the magnetic flux-tubes that are described in Chap. 5 (page 123). It is also supported by the observation that the K-line strength is noticeably greater when splitting occurs, as if the material in that part of the chromosphere were condensed into structures that might or might not happen to lie in the line of sight to the secondary during any one eclipse. Monitoring the phenomenon adequately requires observations of both ingress and egress at the same eclipse, though in practice that is only possible for one eclipse in three, i.e., once in 8 years.

The different series of chromospheric K-line profiles, some of which are illustrated above, indicate that the chromosphere of ζ Aur is far from homogeneous and symmetrical. Evidence of inhomogeneities in the high chromosphere was also demonstrated by Baade et al. (1996), whose models of patterns in the wind outflow and scattering in strong resonance lines underscored the strong dependence of derived densities upon the way those properties are handled. Additional support has come from concentrated polarimetric monitoring; Koch and Pfeiffer (1982) isolated a small variable component of polarization that was intrinsic to the star, but having shown that it could not be due to the B star, either to the wake trailing it or to the scattering of its photons in the K-star's wind, they attributed it to "chemical inhomogeneities" in the K-star's chromosphere. In that respect, one might therefore surmise that ζ Aur is a reasonable template for single cool supergiants (but see Sect. 2.9).

It is also important to note that eclipses in ζ Aur occur about 3 months before periastron, so the distance between the two stars is less during egress than during ingress. Owing to the eccentricity and orientation of the orbit, that reduction in separation corresponds to an increase of at least 40% in the intensity of the B-star's flux at the surface of the K star during egress, compared to ingress. If the ionizing effect of that flux modifies the strengths of the chromospheric features measured during those phases, it will therefore do so differentially.

There is direct evidence that irradiation of the chromosphere by the hot star's flux can alter the state of the chromospheric material in a binary in the region of the sub-stellar point if the two stars are close enough together and if the hot companion is earlier than ~A0. In ζ Aur, phase-dependent emission in Si I at λ 3905 and 4102 Å (and also very weakly in a few lines of Fe I and Fe II) can be observed whenever the B-type secondary is in front of the primary; in ζ Aur that phase commences only 4 weeks after periastron. The emission rises rapidly to a maximum as soon as the hot star begins to irradiate the visible side of the supergiant's disk, and declines as the sub-stellar point moves across to the other limb, whereupon it vanishes. A comparable manifestation of emission in Si I has also been detected in 32 Cyg, though it is much weaker because the inter-star separation happens to be significantly greater at the key phases.

This emission can be understood as a result of photoionization of Si I in the layers where the deep chromosphere borders on the upper photosphere, followed by recombination into the singlet (λ 3905) and triplet (4102 Å) systems. The singlet and triplet cascades that follow radiative recombination lead to emission in strong transitions whose lower levels are close to the ground state, and subsequently undergo multiple scattering. The effect is to produce resonance pumping of the upper levels of the weaker λ 3905 and 4102 Å lines, and leads to their enhanced emission. The efficiency of this conversion from the B-star's UV continuum to Si I line emission is also controlled by Rayleigh scattering by neutral hydrogen in the atmosphere of the K-type primary. That process provides a short-wavelength cut-off to the UV flux which is able to reach the deep chromosphere. In effect, the B-star's UV continuum becomes channelled into the two Si I transitions, leading to the emission that has been detected on the illuminated hemisphere of the K star. The changes in intensity that are observed can be explained as a purely geometrical effect; since the eccentricity of the orbit is nearly 0.4, periastron distance is less than half apastron distance.

2.4.1.2 UV Spectroscopy from Space

During its 1979–1980 eclipse ζ Aur was observed extensively in the UV by *IUE*, and was the first member of the class to be thus monitored spectroscopically from space. The seminal work in the field by Chapman (1981) and by Hack (1981) focused on the prominent characteristics of the spectra that can be attributed severally to the circumstellar envelope, the wind from the supergiant, and the interaction of the B star in the wind. The Mg II lines at λ 2800 Å were unique in exhibiting P Cygni profiles from the expanding wind (Fig. 2.6, left). Broad features in C IV, Si IV and N V were interpreted as arising from plasma at $\sim 10^5$ K, and were consistent with a bow shock and a trailing accretion wake as the B star moved supersonically through

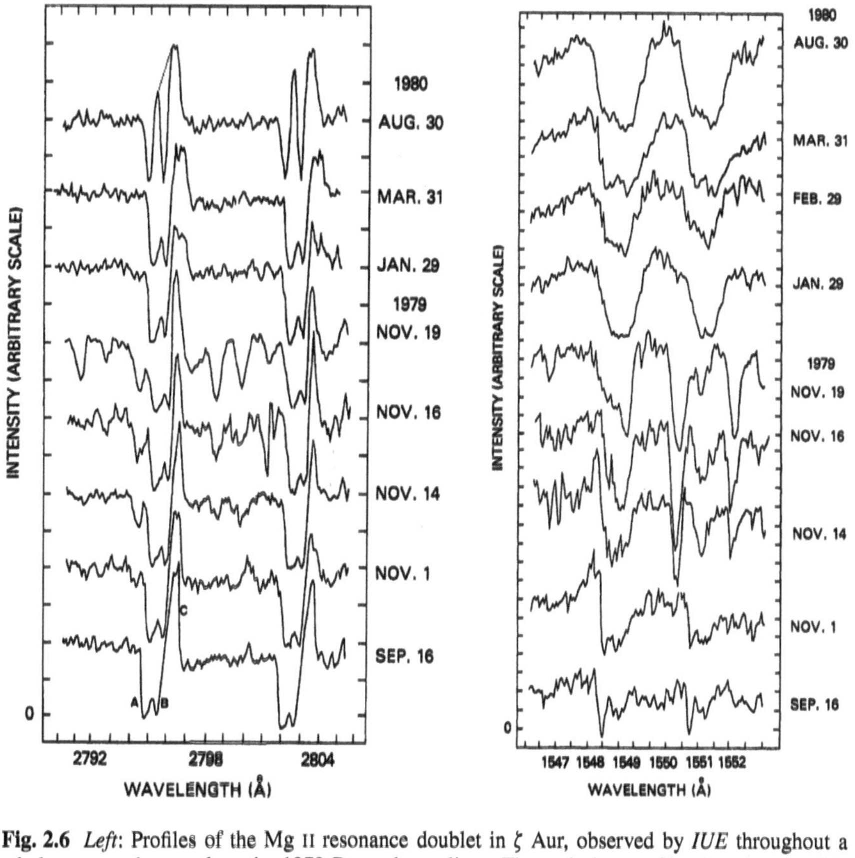

Fig. 2.6 *Left*: Profiles of the Mg II resonance doublet in ζ Aur, observed by *IUE* throughout a whole year, and centred on the 1979 December eclipse. The emission profile altered noticeably after the eclipse. *A*, *B* and *C* identify respectively the components of circumstellar absorption, interstellar absorption and emission from scattering in the wind. *Right*: Profiles of the C IV lines at λ1550 Å. They had grown to maximum strength by August 1980, which is close to secondary eclipse, indicating the presence of absorbing plasma on the side of the B star away from the K star (*Reproduced, with kind permission, from Chapman (1981)*)

the wind (Fig. 2.6, right). From observations of C IV and Si IV, Che-Bohnenstengel and Reimers (1986) modelled the accretion environment near the B star. They found evidence for a temperature gradient in the bow shock: the hotter C IV lines were eclipsed more rapidly than that of Si IV, so the C IV lines are therefore formed closer to apex of the shock cone than the Si IV ones. They also discovered that ζ Aur exhibits a rotationally-broadened C IV emission feature, and attributed it to the formation of an accretion disk of wind material around the hot star. Further observations of Mg II revealed a red-shifted absorption component to the P Cygni profile of the line, and enabled the accretion cone to be mapped (Ahmad et al. 1983). More detailed descriptions and discussions of the early *IUE* observations can be found in the reviews by Hack and Stickland (1987) and by Dupree and Reimers (1987).

Schröder (1985) took those initial *IUE* observations, and also ones for 31 Cyg and 32 Cyg, and constructed curves of growth from relatively unblended absorption lines that were superimposed on the B-star continuum and were not affected by emission components from the wind. Column densities were converted into a radial density distribution through a power-law model. The derived chromospheric density was found to be steeper in ζ Aur than in 31 Cyg or 32 Cyg, but—as Kirsch and Baade (1994) and Baade et al. (1996) later confirmed—the mass-loss rate implied unreasonable wind acceleration and outflow velocities. Eaton (1993a) merged *IUE* data from the 1979 and 1985 eclipses with ground-based observations of the 1990 eclipse, and derived column densities for hydrogen and metallic ions, excitation temperatures and turbulence broadening, by comparing observed line profiles with theoretical ones (see also Sect. 2.4.2.2 on 31 Cyg). Column densities from metallic lines and wings of Lyα were found to match the same distribution with scale height, and were consistent with those from Schröder. Eaton concluded that the chromosphere is supported by turbulent pressure or Alfvén waves (see Chap. 5, page 123), and is probably confined by magnetic fields.

Improvements to these techniques came about through the development of 2-D radiative-transfer codes to compute synthetic line profiles. As Chapman (1981) stressed, when modelling the UV lines in a ζ Aur system one must take into account the fact that the source of the photons from the B star is offset from the source of the wind from the K star. That wavelength shift caused by the velocity of the B star, together with the origin of photons from inside the wind rather than at its base, complicates the computation of line profiles. Che et al. (1983) quantified the wind velocities, turbulence and mass-loss rate of ζ Aur through a newly developed 2-D radiative-transfer code (Hempe 1982) that used a generalized Sobolev approximation to model the resonance lines of metallic ions formed by line scattering in the wind of the supergiant. However, while the results were satisfactory for 31 Cyg and 32 Cyg, they were not so for ζ Aur. Kirsch and Baade (1994) reanalyzed the *IUE* data with an integral-operator approach for arbitrary velocity fields (Baade 1990); although lines of Fe II, Ti II and V II were used to create a starting model of the density structure from a modified curve-of-growth technique, the Si II λ 1533 Å line proved to be more suitable for line-profile

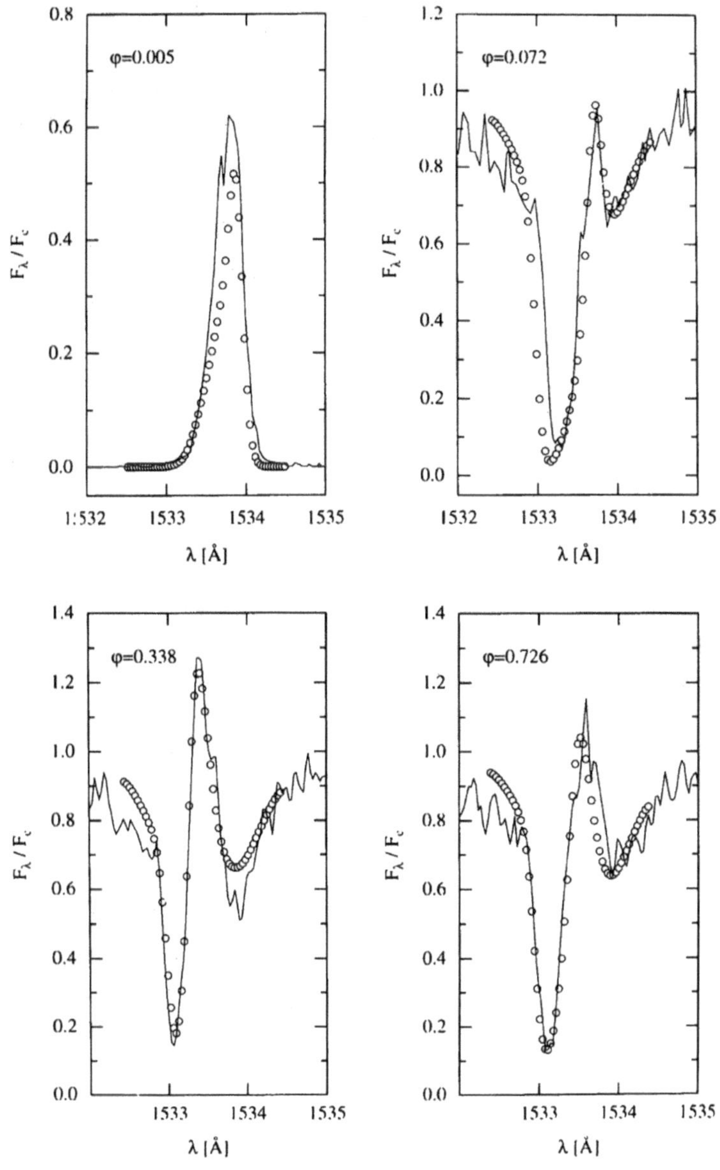

Fig. 2.7 Comparing *IUE* observations of Si II lines at λ 1533 Å in ζ Aur (*full lines*) with theoretical profiles from a 2-D radiative-transfer model (*open circles*) at different orbital phases: totality (*upper left*), secondary eclipse (*lower left*), and quadrature (*upper and low right*) (*Reproduced, with kind permission, from Kirsch and Baade (1994)*)

modelling to determine wind parameters at all phases. Figure 2.7 illustrates the success at fitting its profile as observed at different locations of the B star in its orbit.

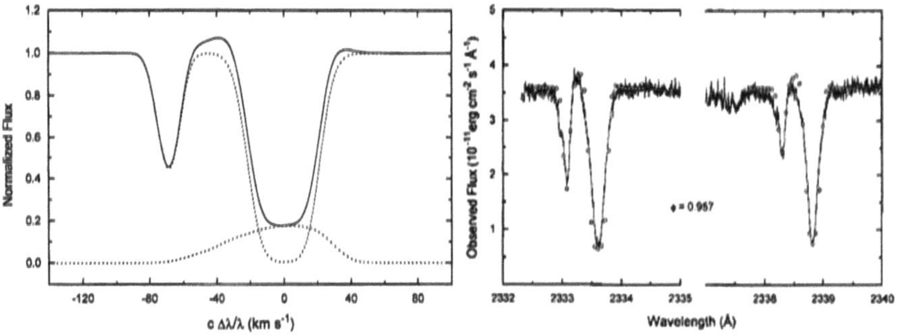

Fig. 2.8 *Left*: Components contributing to a 2-D radiative-transfer model of the Fe II line at λ 2333 Å in the upper chromosphere of ζ Aur during ingress. *Dotted lines*: separate absorption and scattering components. *Solid line*: composite profile. Normalized flux is plotted against velocity from the line centre, for convenience. *Right*: Comparing GHRS observations of Fe II lines (*line*) and the model (*open circles*) (*Reproduced, with kind permission, from Baade et al. (1996)*)

Baade et al. (1996) provided a more detailed treatment of the wind lines observed with both *IUE* and *HST*/GHRS around the 1993 eclipse. As Kirsch and Baade found, the profiles of the Si II line at λ 1533 Å from the higher-resolution GHRS observations could be replicated well by assuming a spherical wind outflow, although two acceleration zones, one below 5 R_K and one above, provided a better correspondence between the model and the observed profile. The Fe II lines, however, were more problematic, particularly when it came to explaining the line splitting seen on ingress. In order to match the profile of Fe II they created an ad hoc model with an ionization deficiency zone; they postulated that the recombination time of Fe II—after the B star and its H II region had passed through the wind—was long enough that the line of sight to it would pass through a trailing cavity of Fe II deficiency, causing an absence of absorption at the velocity sampled there. In effect, one component of the profile came from distant wind, the other from close to the K star. Figure 2.8 shows the model components and the final fit to the observations in the upper chromosphere. It was only when those advances in modelling were applied to the new high-resolution spectra from space that such details in line profiles could be interpreted and inter-compared (Sect. 2.6.1), and the status of this cool primary as a template for a single supergiant thoroughly examined (Sect. 2.9).

2.4.2 31 Cygni

31 Cyg has a period of 10.4 years; Griffin (2008) determines it as 3784.3 days, and also gives a comprehensive account of the history of observing the star at optical wavelengths. The onset of ingress to the very end of egress occupies nearly six months, two of which correspond to totality, thus offering ample opportunity

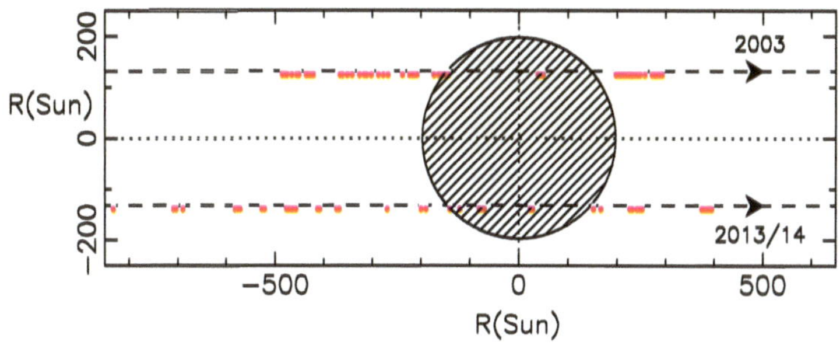

Fig. 2.9 Scaled geometry of two eclipses in 31 Cyg. $R_K = 197\,R_\odot$ (Eaton and Bell 1994) was adopted. For the sake of clarity, the trajectories have been located in the opposite 'hemisphere' of the primary's disk (All the observations represented there were made at the DAO by Griffin)

for observations at different phases even though there are reduced opportunities to monitor many different eclipses.

Figure 2.9 depicts the projected distances of the secondary component relative to the primary's disk corresponding to the phases of the K-line observations that were made during 2003 and 2013–2014 by high-resolution CCD spectroscopy. Eaton (1993c) classified the B star as B3–B4, measured the terminal velocity of the K star's wind as $73 \pm 2\,\mathrm{km\,s^{-1}}$ (but possibly as high as $90\,\mathrm{km\,s^{-1}}$), and determined a mass ratio for the system of 1.66 ± 0.03, with individual masses of $11.7\,M_\odot$ (K star) and $7.1\,M_\odot$ (B star). The relatively large actual separation between the component stars should mean that the spectrum of 31 Cyg shows much weaker binary interactions in the UV than do those of either ζ Aur or 32 Cyg, so its chromospheric absorption spectrum might be more straightforward to analyze; even so, the many specific oddities described and illustrated by Wright and Odgers (1962) and by Wright (1970) have continued to inspire research interests in 31 Cyg.

2.4.2.1 Ground-Based Monitoring

As explained by Griffin (2007), a series of optical CCD spectra of the secondary component, extracted from observations of the 2003 eclipse, revealed changes in the chromospheric K-line profile during both ingress and egress (Fig. 2.10, left). Just as Wright (1970 and references cited therein) found in earlier eclipses, during ingress the line showed multiple components that became more symmetrical and centred at the K-star's rest wavelength as the eclipse deepened. Egress (particularly in 2003) also indicated considerable structure in both persistent and in more transient components.

The 'satellite' lines in Fig. 2.10 exhibit little systematic or inter-connected behaviour either from ingress to egress or from one eclipse to another. Some are confused by blending between components moving at different velocities; their RV displacements are occasionally quite high (as on days −43 and −42). As with ζ Aur, there is a tendency for increased complexity during egress compared to ingress,

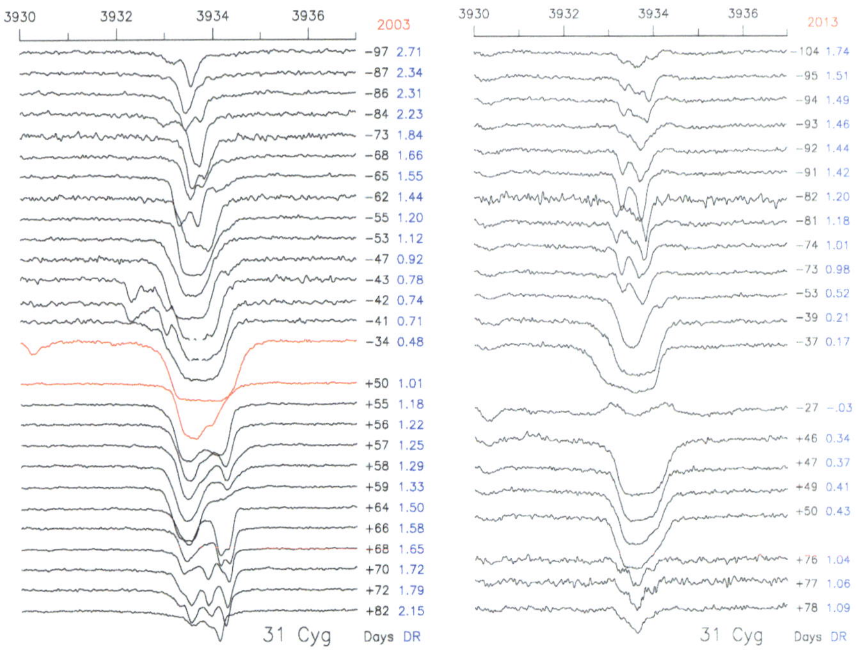

Fig. 2.10 Chromospheric K-line profiles in 31 Cyg during ingress and egress in 2003 (*left*), and in 2013–2014 (*right*). 'Days' and DR are as for Fig. 2.3. The spectra from 2003 are of the B-star alone, plus chromospheric absorption. In contrast, the spectra from 2013 are of the composite system as observed, in order to avoid contamination by the emission that is evident in the giant itself (day -27); the contribution from the giant star is in any case quite small at these wavelengths (Data source: CCD spectra observed at the DAO by Griffin)

but while there seemed to be an established red-shifted satellite feature during egress in the 2003 eclipse with a fairly steady velocity of about $+45\,\mathrm{km\,s^{-1}}$, it mysteriously vanished for at least 1 day between days $+59$ and $+64$. The lack of a chromospheric signature at $+82$ days in 2003 does not represent a physical boundary to the chromosphere since chromospheric K-line absorption was visible at greater heights in ingress in both eclipses shown there, though it might instead be a signal that Ca was doubly ionized. In 2013 the K line exhibited splitting at day -91 that was similar to that detected in ζ Aur, but it probably did not reflect the same cause as it lacked both the persistence and the precise restriction in component velocities that characterized the situation so specifically in ζ Aur. Egress from the 2013 eclipse was difficult to follow for long because of unfavourable dates, and the next, centred in 2024 March, will be no better. It is not until 2034 July that complete monitoring of the majestic changes seen in the chromospheric eclipses of this binary will again be feasible.

2.4.2.2 UV Spectroscopy from Space

The 1982 eclipse of 31 Cyg was the first such event to be observed for this system with *IUE* (Stencel et al. 1984). The extensive eclipse photometry which was also obtained supported an accurate refinement of both the timings of ingress and egress and the photometric period. The *IUE* spectra revealed the expected rich array of emission lines during totality arising from radiation scattered in the primary's extended expanding atmosphere, and showed how the same emission filled in parts of the chromospheric absorption lines that were so prolific during atmospheric eclipse (Fig. 2.11).

Bauer and Stencel (1989) subsequently produced an atlas of the UV spectra, listing the emission lines seen in total eclipse and absorption lines seen at atmospheric-eclipse phases. *IUE* observations made during the 1992–1993 eclipse covered both ingress and egress; Eaton and Bell's (1994) analysis of them yielded the physical properties of the gas throughout the extended atmosphere, and a model of the distribution of the gas that assumed only radial dependence and large-scale homogeneity (Eaton 2008). Neutral hydrogen column densities, determined from the increase in the red wing of Lyα during eclipse (Fig. 2.12), agreed within the errors of measurement with those calculated from the metallic lines. Evidence that hydrogen is ionized in a shell around the B star, and that material is collapsing onto the B star at seemingly random phases, introduced new modelling constraints.

The chromospheric lines in 31 Cyg are broader than those in ζ Aur, and one contributor to that broadening is the differential expansion of the upper chromosphere. By mapping radial expansion in a wind onto the line of sight, Eaton (1996) demonstrated how such an expansion can model the observed line profiles well (Fig. 2.13, upper panel) even though the actual contribution from random

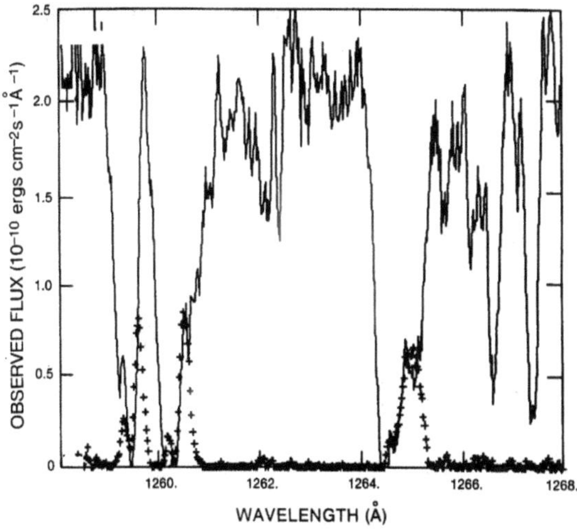

Fig. 2.11 This UV region of 31 Cyg, observed by *IUE*, shows only emission lines during mid-eclipse (*plusses*). They are displaced redward compared to the strong absorption features seen before eclipse (*solid line*) (*Reproduced, with kind permission, from Stencel et al. (1984)*)

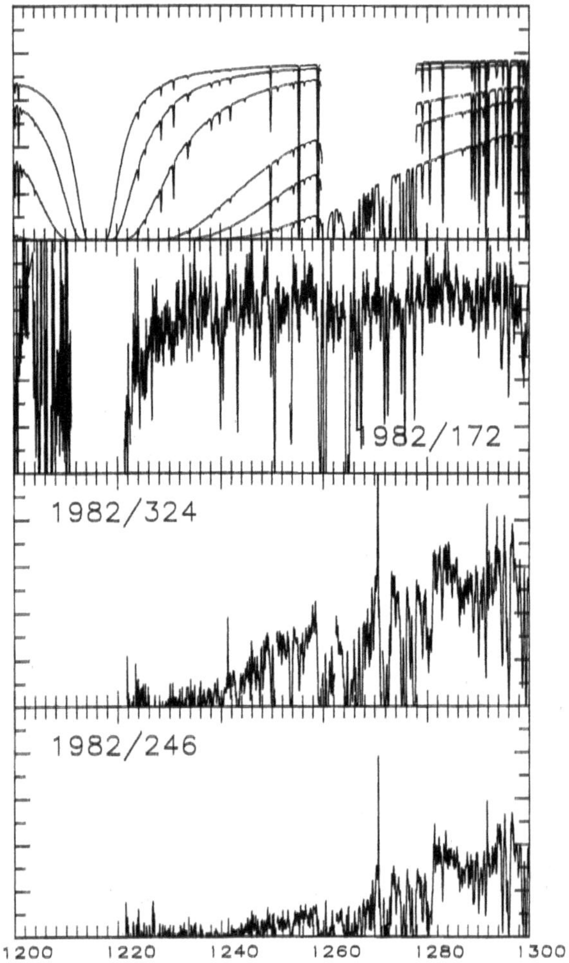

Fig. 2.12 Calculated and observed changes in the Lyα profile during chromospheric eclipse. *Top panel*: Broadening of the wings of Lyα in a sequence of mass column densities 0.001, 0.003, 0.01, 0.05, 0.10 and 0.20 g cm^{-2}, respectively, increasing downwards. *Lower panels*: *IUE* spectra of 31 Cyg at three epochs of the 1982 eclipse, ranging from the onset of chromospheric eclipse (*upper*) to deep chromospheric eclipse (*lower two*). The original spectra have been divided by a spectrum observed far from eclipse in order to minimize the contributions from the B-star's photosphere (*Reproduced, with kind permission, from Eaton (1988)*)

motions can be no more than 15 km s^{-1} and is most likely in the range 7–10 km s^{-1}. The lower panel of Fig. 2.13 illustrates how the match to observation is impaired when a single high turbulence is adopted.

Like ζ Aur, 31 Cyg too exhibits a shock cone caused by the motion of the B star in the wind of the supergiant, but it does not show an accretion disk although broad absorption features near inferior conjunction of the B star are indicative of an accretion wake (Che-Bohnenstengel and Reimers 1986).

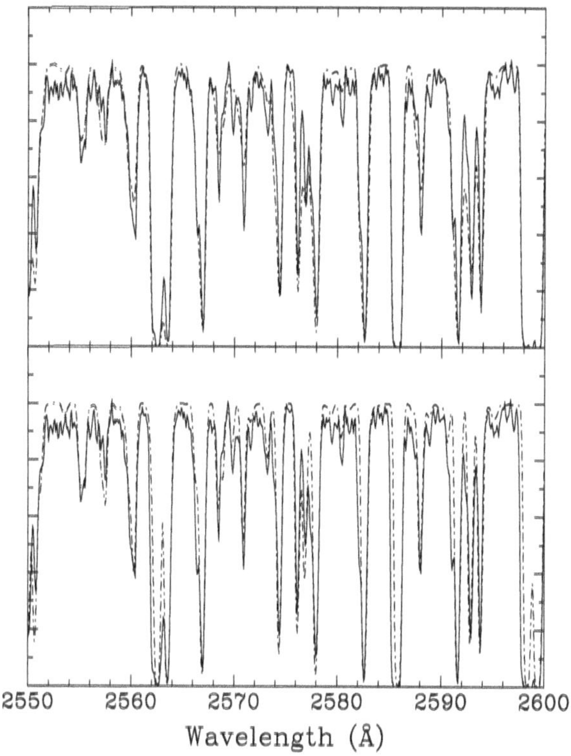

Fig. 2.13 *Upper*: A shell spectrum of 31 Cyg at phase .976 (*solid line*) compared to a synthetic spectrum calculated from a differentially expanding wind model incorporating a wind velocity of 7 km s^{-1} and a projected expansion of −83 to +41 km s^{-1}. *Lower*: The same observation compared to a single-component model incorporating a 'turbulence' velocity of 22 km s^{-1} (*Reproduced, with kind permission, from Eaton (1996)*)

During its 2003 eclipse 31 Cyg was also observed by *FUSE*, once in deep chromospheric ingress, twice during total eclipse and twice during egress—one in deep chromospheric phase and the other in a more shallow one (Bauer and Bennett 2014). During total eclipse the *FUSE* spectrum exhibited emission lines from species similar to those seen with *IUE* at longer wavelengths, the strongest arising from N II; fig. 4 of Stencel et al. (1984) and fig. 10 of Eaton and Bell (1994) show examples. Others came from neutral O and Ar, singly-ionized Fe, P, Si and S, and from double-ionized Fe and S. Some were blended with components of interstellar absorption. Emission lines from the singly-ionized ions were approximately symmetrical and at velocities near that of the system, while more highly ionized lines were asymmetrical (steeper on the long-wavelength side) and were red-shifted from the systemic velocity. That evidence suggested a velocity gradient in the upper layers, and scattering by a relatively symmetrical wind that had a zone of higher ionization (Fe II → Fe III) nearer the B star, as proposed by Eaton and Bell (1994,

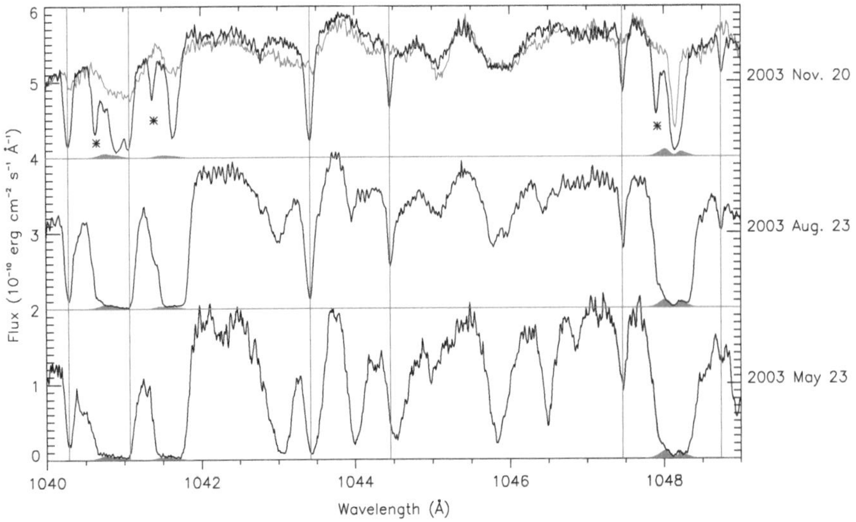

Fig. 2.14 A sequence of *FUSE* spectra of 31 Cygni observed in 2003: deep chromospheric eclipse in ingress (*bottom*), mid-chromospheric eclipse in egress (*middle*), and nearly out of eclipse (*top*). The spectra are plotted on the same vertical scale; the *horizontal grey line* marks the zero level for the two upper panels. The total eclipse spectrum (mid-eclipse 2003 June 30) is shown as *solid grey*. The *FUSE* spectrum of the B5 V star HD 36981, drawn in *grey*, has been included in the top panel for reference, and has been scaled in both flux and wavelength to represent the likely contribution of the B-star spectrum to that of 31 Cyg. *Vertical grey lines* indicate absorption by interstellar H_2. *Asterisks* mark wind absorption, seen here about $-80\,\mathrm{km\,s^{-1}}$ shortward of strong features of O I and Ar I. The core profiles of the stronger chromospheric absorption lines are modified by the emission which occurs in the same features during total eclipse

their fig. 11). Well into egress the sharp blue edge of the O I line showed saturated absorption extending over most of the velocity range of the wind; in fig. 10 of Eaton and Bell (1994) that effect is modelled for O I.

During deep chromospheric eclipse the *FUSE* observations showed rich absorption spectra of neutral and singly-ionized species (Fig. 2.14). The cores of some of the deepest absorption features, in particular of N II, were slightly filled in by emission from the wind. Line splitting that seemed to parallel what was seen in the Ca II *K* line of ζ Aur (Figs. 2.4 and 2.5) was also observed in several of the strong features in egress several weeks beyond 4th contact, but—as with the optical spectra of ζ Aur—not in ingress.

2.4.2.3 Quasi-periodic RV Variations

Systematic RV observations of the cool giant of this system by Griffin (2008), and also of 32 Cyg and ζ Aur, have brought to light unexpected deviations, of the order of $1\,\mathrm{km\,s^{-1}}$, from strict orbital motion. In Fig. 2.15 (for which the data were acquired

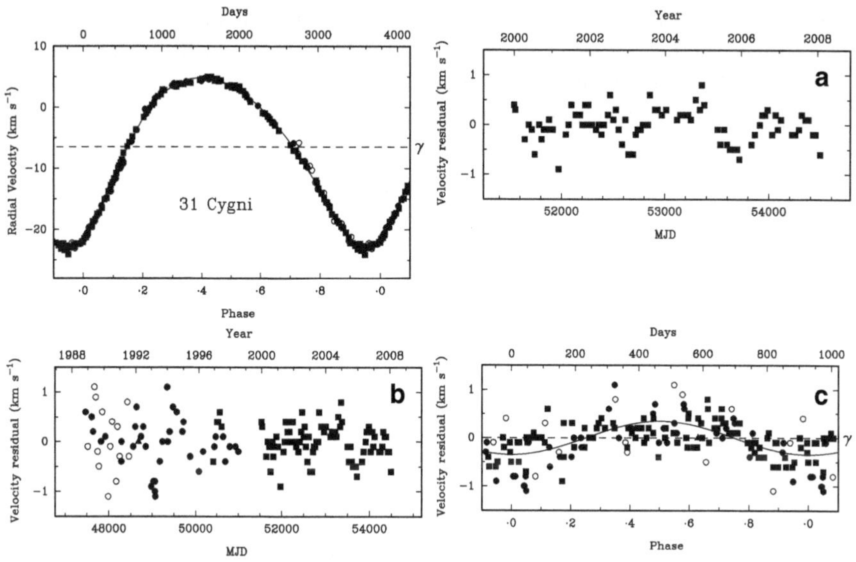

Fig. 2.15 *Upper left*: The RV orbit of 31 Cyg, determined by R.F. Griffin from measurements made with the original Cambridge (UK) RV spectrometer (○), the Cambridge *Coravel* (■) and the OHP *Coravel* (●). Panel (**a**) plots the residuals as a function of time during an interval of intense observing, (**b**) plots them over a much longer interval, while in (**c**) they have been folded about an apparent period of 929 ± 9 days (*Reproduced, by kind permission, from Griffin (2008)*)

exclusively by *Coravel*-type RV spectrometers and are therefore of high precision), an examination of the residuals from the RV orbit of the primary (upper left) reveals a periodicity (measured as 929 ± 9 days) that was particularly evident during a limited interval when the system was observed systematically (panel **a**). The same periodicity persisted for a much longer interval (panel **b**) in the early years, though with increased amplitude, and was probably of not the same wave-form. In panel **c** the residuals from the whole set have been folded about the 929-day period; the double maximum which is suggested in panel **a** is now seen more clearly. Consistent independent evidence, reported by Eaton et al. (2008), confirm the origin as stellar and not instrumental; at least one prominent dip in the residuals appeared in both data sets. Similar RV observations of 32 Cyg by Griffin brought to light a transient periodicity in its residuals too, of 775 ± 27 days, while Eaton et al. found cyclic residuals for ζ Aur as well, though they did not derive a period for any of those stars.

Griffin interpreted the patterns as a signature of the rotation of surface irregularities, loosely called ‘spots’. Eaton et al. argued that they cannot be caused by pulsations because the amplitudes and durations of the velocity excursions would imply unrealistically large changes in radius, and suggested that the velocity excursions more probably represent stochastic variations in the large-scale convective flows in the atmospheres of these supergiants. The effects may be manifestations of the 'star-patches' that have been found in a number of cool stars (e.g., Toner and Gray 1988; Gray 2008), or the large-scale convection reported by Kiss et al. (2006). Correlation between the observed periodicity and the rotation of the supergiant in 31 Cyg may be possible, since a 929-day period corresponds to an equatorial velocity of the order of 8–10 km s^{-1}, which is close to the value of 6.8 km s^{-1} derived from the *Coravel* RV 'dip' after an appropriate correction for turbulence.

2.4.3 32 Cygni

The K5 Ib + B6 V system of 32 Cyg has a period of 1147.51 ± 0.31 days (Griffin 2008), so there are more frequent opportunities to monitor its eclipses than in the case of 31 Cyg. Griffin (2008) includes a comprehensive history of observing efforts. A precise mass ratio for 32 Cyg is just as elusive as it is for 31 Cyg, and for the same reasons; Wright's (1970) values of $M_K = 9.7\,M_\odot$ and $M_B = 4.8\,M_\odot$ for the component masses have not yet undergone much revision. An interesting complication arises through an ambiguity in the radius of the supergiant. According to Eaton et al. (2008), an ellipsoidal variation which has been detected in the photometric light curve cannot be fitted if the supergiant's radius is as small as the 175 $R_\odot$ adopted by Eaton (1993b)—a conclusion also reached earlier by Guinan and McCook (1979)—but if the radius were even as high as 260 $R_\odot$, which is still short of the 300 $R_\odot$ or so required to fit the photometry, it would imply a smaller inclination and somewhat higher masses to fit the photometric observations. For the purposes of Figs. 2.16 and 2.17 a value for R_K of 175 $R_\odot$ was adopted.

32 Cyg exhibits only short eclipses—totality lasts for about 1 day in the near UV but is scarcely reached at visible wavelengths—and are assumed to be grazing. The projected trajectory of the hot secondary therefore samples polar regions of the chromosphere, as opposed to lower-latitude ones as in 31 Cyg, and the more extreme characteristics and changes that are revealed in its chromospheric absorption-line profiles seem to reflect that different geometry: the lines are broader, indicating greater turbulence, satellite lines behave in a more random manner and with higher velocities, and there is altogether greater variety from one eclipse to another.

2.4.3.1 Ground-Based Monitoring

Because eclipses in 32 Cyg are three times more frequent than in 31 Cyg, correspondingly more eclipse-related spectra have been acquired even though totality is so brief. The four series of K-line spectra, derived from eclipse observations obtained between 2003 and 2012 (Fig. 2.17), display interesting details of the inhomogeneity and turbulence just mentioned. Figure 2.16, a scaled geometrical model of the system, shows the projected positions of the two components during the four eclipses when the spectra in Fig. 2.17 were obtained.

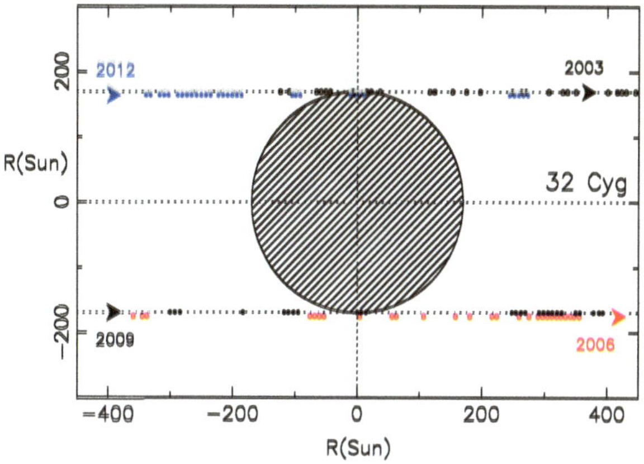

Fig. 2.16 Scaled geometry of eclipses in 32 Cyg. For the sake of clarity, the trajectories of two of the eclipses have been located in the opposite 'hemisphere' of the primary's disk (Data source: CCD spectra, observed at the DAO by Griffin)

A point of interest in Fig. 2.17 that must not be missed is the steady decrease in the degree of complexity and randomness of individual satellite lines, proceeding from a haphazard appearance in 2003 to a very orderly one in 2012 that is almost devoid of extraneous features. An analogy might be drawn with the case of the Sun: during sunspot minimum the solar corona (and presumably the chromosphere with it) has broad, smooth structures, while at sunspot maximum it has a smaller-scale, spiky appearance. 32 Cyg may therefore be exhibiting direct evidence of magnetic cycles, which do indeed occur in other cool stars. The presence of some level of magnetic activity in the cool supergiant is also evident in changes in the *K*-line core profile observed in different total-eclipse spectra; emission reversals were present, and in almost equal strength, in totality spectra observed in 2003 and 2012, but in 2009 the violet component had vanished. A series of *K*-line spectra during the next eclipse (in late October 2015) could contribute important constraints concerning the nature and cause of this intriguing variability, though it may be difficult to monitor the system for as long as 3 months during egress.

2.4.3.2 UV Spectroscopy from Space

The first spectra of a ζ Aur system taken with *IUE* were back-to-back observations of 31 Cyg and 32 Cyg. Even the initial observation of 32 Cyg demonstrated that this star exhibits a fascinating spectrum in the satellite UV (Stencel et al. 1979). Although the phase observed was far from eclipse, a multitude of intense P Cygni emission features appeared in numerous lines superimposed on the B-star continuum (Fig. 2.18), indicating that the B star must be embedded in the wind of the K star. Stronger P Cygni lines, such as Mg II λ 2800 Å, O I λ 1302 Å and numerous Fe II features, exhibited a complex of multiple absorption components

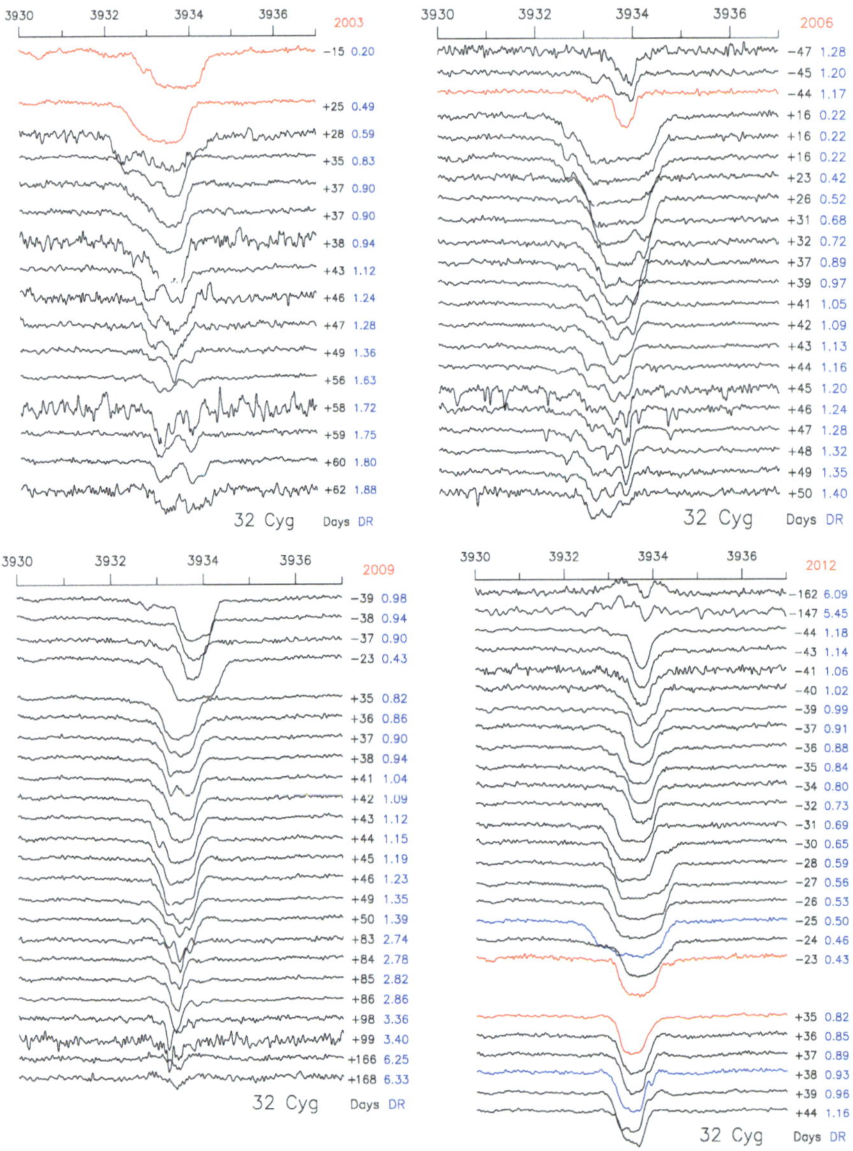

Fig. 2.17 Chromospheric *K*-line profiles in 32 Cyg during ingress and egress in 2003, 2006, 2009 and 2012. The spectra are of the B-star alone, plus chromospheric absorption. 'Days' and DT are as for Fig. 2.3 (Data source: CCD spectra observed at the DAO by Griffin)

ranging beyond $-200\,\mathrm{km\,s^{-1}}$ from rest velocity. Later, emission lines of C IV and Si IV were seen during eclipse, indicative of a bow shock, but in contrast to other ζ Aur systems the lines did not vary during partial eclipse phases, and there seemed to be no accretion wake (Che-Bohnenstengel and Reimers 1986). Differences of

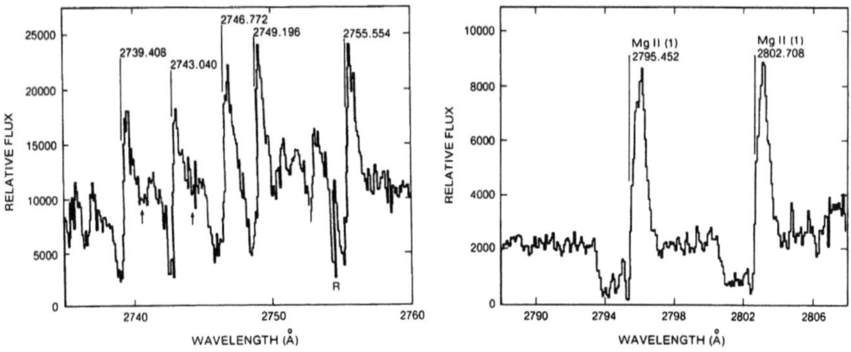

Fig. 2.18 *IUE* observations of 32 Cyg far from eclipse, showing P Cygni profiles in lines of Fe II (*left*) and Mg II (*right*). *Arrows* indicate additional high-velocity absorption components in Fe II; 'R' is a camera réseau mark (*Reproduced, with kind permission, from Stencel et al. (1979)*)

that genre maybe attributable to the grazing nature of the eclipse, since an extended shock cone could be observable above the limb of the K supergiant even during an eclipse of the hot star itself. Observations at other phases revealed that the P Cygni emission lasts throughout much of the orbit.

Schröder (1985) studied the upper chromospheric density distribution and wind acceleration of 32 Cyg from 11 *IUE* spectra observed at phases near the 1981 eclipse. A curve-of-growth technique was applied to unblended chromospheric lines of ionized Fe, Ti and V. A model for continuous wind outflow proved to be a good average approximation, though Baade et al. (2001) uncovered discrete condensations in outflowing material. In the same eclipse, extra Rayleigh scattering by H I at $\sim\frac{1}{6}\,R_K$ above the limb of the star suggested a stellar prominence (Schröder 1983); some unsaturated lines also became asymmetrical at that time through additional absorption at $\sim$20 km s^{-1}, in keeping with the characteristics of a quiescent solar prominence. In a separate study of the lower chromosphere of 32 Cyg, Schröder (1986) used five pseudo-continuum regions from λ 1350 to 2992 Å to ascertain changes in continuum opacity during deep eclipse (Fig. 2.19), and was thence able to demonstrate that Rayleigh scattering at the wings of Lyα was the dominant scattering process. The trend also indicated how the duration of eclipses in the UV lengthened with decreasing wavelength—a commonality in all the ζ Aur systems studied so far.

Analyses by Eaton (1993b) of optical and archived *IUE* spectra have highlighted various differences in the outer regions of the primary of 32 Cyg compared to ζ Aur or 31 Cyg. Eaton et al. (2008) deduced an increase with height in both turbulence and excitation temperature (the latter rising to 12,000–17,000 K), and suggested that magnetic fields could cause a high degree of clumping of the gas.

Of the three classical ζ Aur systems, 32 Cyg has the strongest binary effects as judged from the strength of wind-scattered lines in total eclipse. It may thus represent a step in the progression from the low interaction of 31 Cyg to the increasingly energetic ones of VV Cep and symbiotic binaries in which immersion

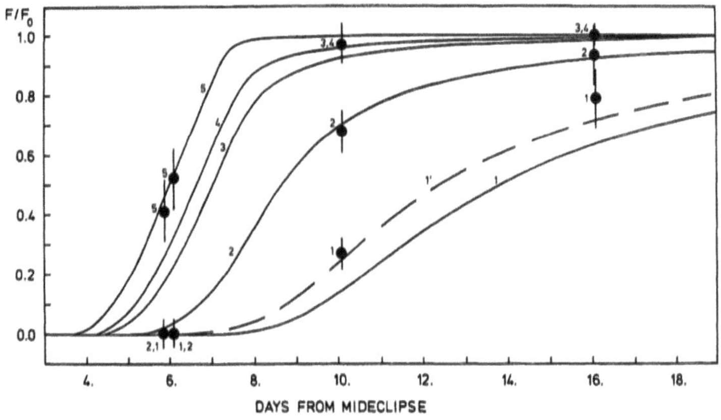

Fig. 2.19 Partial eclipse measurements of 32 Cyg at pseudo-continuum points (*1*) λ 1350, (*2*) 1513, (*3*) 1783, (*4*) 1960 and (*5*) 2992 Å. The curves are continuum fluxes transmitted by a Rayleigh scattering model (*Reproduced, with kind permission, from Schröder (1986)*)

of a hot object in a massive wind dominates the appearance of the spectra (Kenyon 1986). High-quality polarimetric observations of 32 Cyg made during the eclipse of 1993 October (Fox and Griscom 1996) revealed variability that was inconsistent with just scattering from the atmosphere of the supergiant; rather, it called for additional sources, such as sporadic mass transfer into an accretion disk around the B star together with some degree of dispersion of that disk. That concept of sporadic mass transfer has also been deduced from the evidence from UV spectra of VV Cep, as explained in Sect. 3.8.

2.4.3.3 Quasi-periodic RV Variations

As mentioned in Sect. 2.4.2.3, the quasi-periodic patterns that were evident in the residuals of the RV orbit for 31 Cyg (Fig. 2.15) were also detected in 32 Cyg (Griffin 2008; Eaton et al. 2008), but with a periodicity of 775 ± 27 days compared to 929 ± 9 days for 31 Cyg.

Griffin drew another interesting inference about the systems of 31 Cyg and 32 Cyg. The *Hipparcos* proper motions, and the *Tycho-2* ones that take into account positional measurements over a very long time-base, are identical within their uncertainties, and the orbital systemic velocities of the two systems differ by only $0.032 \pm 0.04\,\mathrm{km\,s^{-1}}$. Although curiously close together in the sky, the possibility that they could be a double star seems not to have been considered. Their projected separation of about 5 parsecs (10^6 AU) may seem too great for them now to constitute a gravitationally bound pair, though Dhital et al. (2013) have already identified wide pairs of low-mass stars with separations of 10^3–$10^{5.5}$ AU in the Sloan Digital Sky Survey data on the basis of similar criteria.

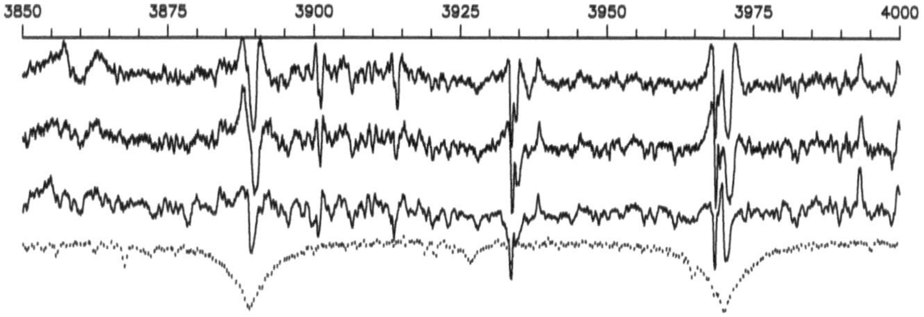

Fig. 2.20 Optical spectra of VV Cep are dominated by the hot emission-line component even near Ca II *H* and *K*. The *dotted spectrum* is that of the standard B2 V star 51 Cyg (Data source: Calar Alto and Mount Wilson photographic spectra obtained by Griffin)

2.4.4 VV Cep

The VV Cep class refers to binaries containing an M-type supergiant and a B-type secondary. Only VV Cep itself is known to exhibit eclipses, so it alone is included in the ζ Aur class too. The three spectra in Fig. 2.20, observed on occasions far from eclipse, illustrate the highly complex nature of its composite spectrum. As the comparison with the spectrum of the B2 V standard 51 Cyg (drawn dotted) shows, the hot secondary in VV Cep appears to be embedded in circumstellar gas at all phases and its photospheric spectrum cannot be seen even in this optical region. The complexity of this system has earned it a chapter to itself (Chap. 3, page 85).

2.4.5 22 Vul

The discovery of eclipses in the SB1 system 22 Vul was one of the early successes of *IUE*. This binary, which appeared to contain a cool supergiant and a hot but much fainter dwarf, was exactly the sort of system whose properties could only be elucidated through UV spectroscopy. However, the first *IUE* observations did not show a simple composite spectrum, but rather the signatures of a dense, high-velocity wind from the primary star. That led to the suspicion that it could be a warm analogue of the classical ζ Aur systems. Follow-up spectroscopy and photometry provided not only further incontrovertible evidence of a late-B component, but also that it underwent eclipses by the G-type supergiant primary for about 8 days (Parsons and Ake 1983; Ake et al. 1985). The measurements also supported the spectral classification of G7 Ib–II + B8.5 V that is now adopted (Griffin et al. 1993; Parsons and Ake 1998). The orbital period of 249.131 ± 0.006 days (Griffin et al. 1993) honours this system as having more frequent eclipses than the classical ζ Aur stars, though the large luminosity difference between the primary and the secondary also makes it hard to detect weak chromospheric-eclipse features in ground-based spectra.

22 Vul thus extended the known ζ Aur class to G-type supergiants, and offered opportunities to study ways in which the chromosphere of a supergiant that was considerably warmer than those studied so far might differ from them. Particular interest also attached to 22 Vul because the primary star lay in the region of the H–R diagram that is populated by stars with hot, solar-like coronæ as distinct from the K-type supergiants with extended chromospheres and cool winds. Satellite UV spectra of those warmer giant stars show lines of highly ionized species that are weak in, or absent from, the K supergiants. The profiles of Mg II, Cr II, Mn II and Fe II also contain evidence of a massive wind, while absorption in Fe II extends for at least 2 R_G above the photosphere (Schröder and Reimers 1989; Eaton and Shaw 2007).

2.4.5.1 Ground-Based Monitoring

The challenge of extracting the spectrum of the secondary in the optical near-UV region is made clear in Fig. 2.21. The upper set reproduces the composite spectra as observed during successive nights of ingress (Fig. 2.22), but it is only when the supergiant's own spectrum, observed during total eclipse, has been subtracted away—as in the lower set—that the wealth of information in those chromospheric features, both strong and weak, can be appreciated. The respectable S/N ratios needed for precise measurements of weak chromospheric-eclipse features in the optical region are clearly hard to achieve in a primary-dominated system like this. It is now known that strong absorption in the Ca II K line persists to a height of at least 0.5 R_G, but further monitoring at high S/N is needed in order to assess the full extent of that cool chromosphere. Nevertheless, Schröder et al. (1994) were able to derive an empirical density model for the low chromosphere of 22 Vul by applying the curve-of-growth method to 260 measured lines representing 12 species of atoms and ions, and by modifying the method to analyze the Ca II K line too. Perhaps somewhat surprisingly (and as elaborated in more detail in Sect. 2.6.1), the hydrogen particle densities in 22 Vul proved to be very similar to those at corresponding heights in ζ Aur up to about 9 $R_\odot$ above the photosphere, though at greater heights the densities of easily-ionized species such as Fe I and Ca II decrease much more rapidly in ζ Aur because the flux from its mid-B-type secondary has a greater ionizing effect in its outer chromosphere than in the case for 22 Vul, which has a late-B secondary.

2.4.5.2 UV Spectroscopy from Space

22 Vul was the first ζ Aur system for which low-dispersion spectrophotometry from *IUE* was central in determining the system's parameters and the circumstances of the eclipse (Parsons et al. 1985; Ake et al. 1985). As a complement to Fig. 2.21,

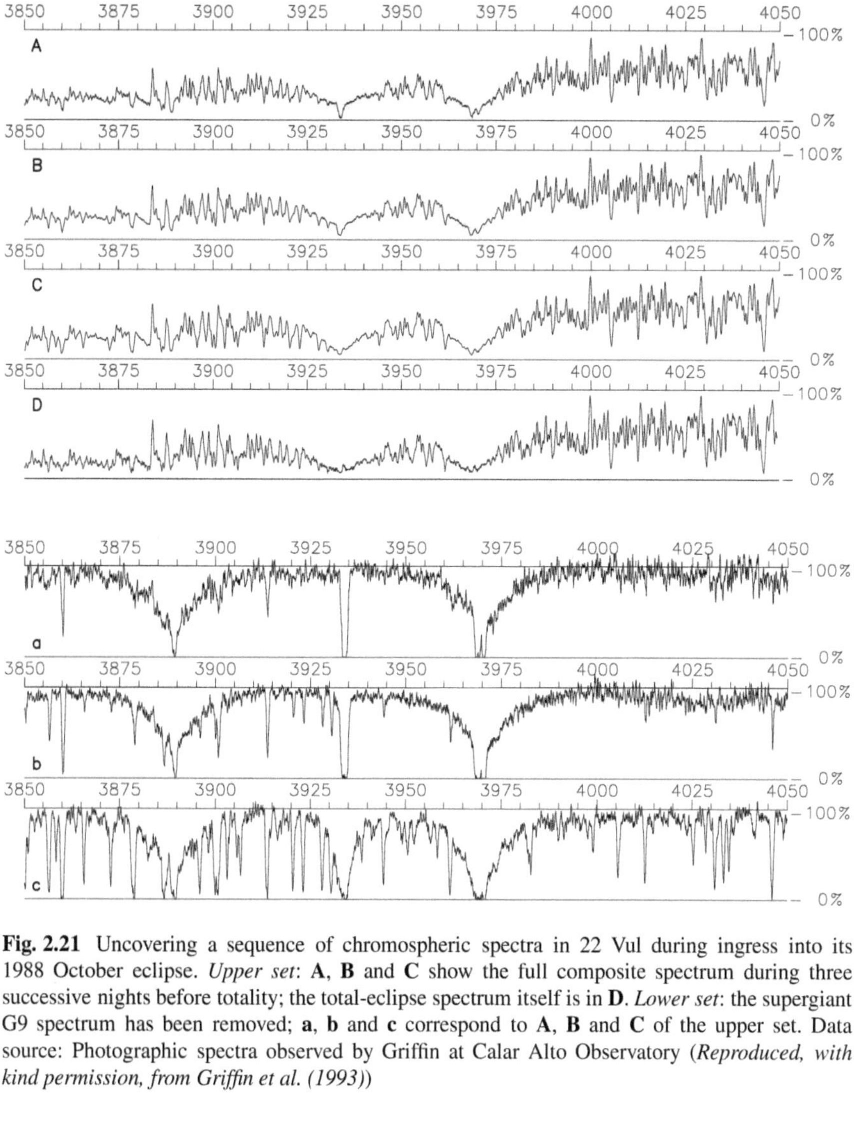

Fig. 2.21 Uncovering a sequence of chromospheric spectra in 22 Vul during ingress into its 1988 October eclipse. *Upper set*: **A**, **B** and **C** show the full composite spectrum during three successive nights before totality; the total-eclipse spectrum itself is in **D**. *Lower set*: the supergiant G9 spectrum has been removed; **a**, **b** and **c** correspond to **A**, **B** and **C** of the upper set. Data source: Photographic spectra observed by Griffin at Calar Alto Observatory (*Reproduced, with kind permission, from Griffin et al. (1993)*)

Fig. 2.23 illustrates spectra in the UV at selected orbital phases. The cross-over where the two stars contribute equal amounts of light occurs near λ 2600 Å, and explains why the spectrum of the secondary is difficult to detect at optical wavelengths. During chromospheric-eclipse phases, regions such as λ 2300–2700 Å, in which the overall profile is defined at low dispersion by the confluence of absorption lines (mainly of low-excitation Fe II), become exaggerated, while other regions (e.g., $\sim \lambda$ 1900–2100 Å) change little until the B star has almost become occulted. At high dispersion, Mg II, Fe II, Cr II and Mn II exhibit P Cygni profiles to some degree and an added wind absorption at $-200\,\mathrm{km\,s^{-1}}$ (Ake et al. 1985). The surface flux of the

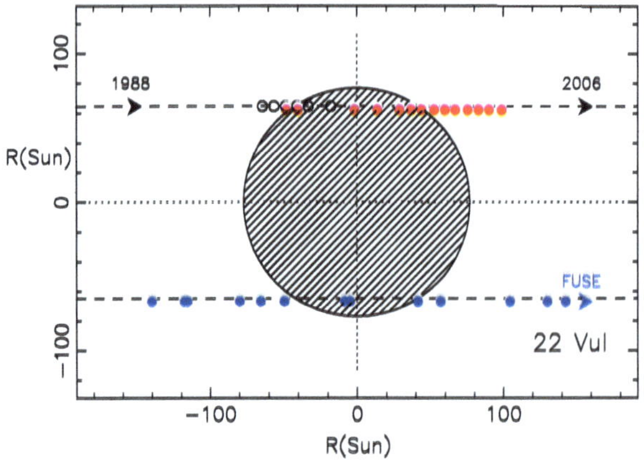

Fig. 2.22 Scaled geometry of eclipses in 22 Vul. $R_K = 77\,R_\odot$ was adopted (Griffin et al. 1993). The upper trajectory represents ground-based observations of eclipses in 1988 (*open circles*) and 2006 (*filled circles*), while the lower one merges the phases of *FUSE* observations made in 2002, 2003 and 2004. The trajectory for the space data has been located in the other 'hemisphere' for clarity

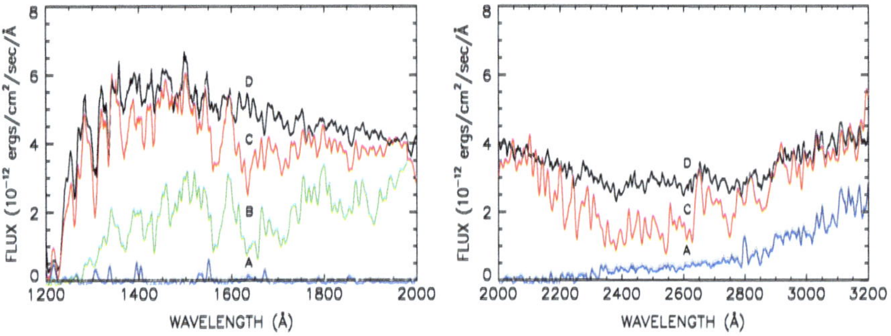

Fig. 2.23 Low-dispersion *IUE* spectra of 22 Vul, showing various orbital phases that range progressively upwards in flux from total eclipse at the bottom (**A**), through partial eclipse (**B** and **C**), to outside eclipse (**D**). At **D** the spectrum is dominated by the B8.5 V secondary at all the short wavelengths, and by the G supergiant at the longest ones; the two equalize near λ 2600 Å. At **A** the spectrum of the hot star has disappeared, and the G-star spectrum is overlain with strong emission lines from Si IV λ 1400, C IV λ 1550 and Al II λ 1670 Å. In **B** and **C** the combined spectrum is fully composite, but includes the type of prominent absorption lines which characterize a chromospheric-eclipse spectrum (*Reproduced, with kind permission, from Ake et al. (1985)*)

emission lines at totality is much larger than has been detected from the transition regions of other late-type supergiants, and was interpreted (Che-Bohnenstengel and Reimers 1986) as contributions from a shock front at the B star, as is also seen in other ζ Aur binaries; the attendant accretion cone was discovered by Ahmad and Parsons (1985) from additional red-shifted absorption in Mg II *h* and *k*.

Compared to ζ Aur, 31 Cyg and 32 Cyg, analyses and modelling of the chromosphere of 22 Vul have been somewhat more problematic. A chromospheric-density model out to $2\,R_G$ was derived by Schröder and Reimers (1989) from a curve of growth of low-excitation absorption lines of Fe I, Fe II, Ti II and V II. Those lines occur at the long-wavelength end of the *IUE* spectra, and are inevitably more contaminated by underlying photospheric lines from the G star than is the case in systems with K-type supergiant primaries. However, they represent the dominant ionization state, primarily through the radiation field of the B star, and they do seem to yield a consistent overall density. Indeed, when the column densities are transformed into radial densities through a power-law model and scaled to stellar radius, the density and its gradient are comparable to those of the K supergiant primaries. Again, as found by Reimers and Che-Bohnenstengel (1986), the turbulence velocity of $\sim$50 km s^{-1} in the outer chromosphere is much higher numerically than for the K supergiants, but the ratio of turbulence velocity to wind velocity is about the same in 22 Vul as in the K supergiants, thereby suggesting a link between the acceleration of the wind and atmospheric turbulence.

The main difficulty in analyzing 22 Vul arises from trying to fit the absorption and emission components simultaneously, particularly those of Fe II, and requires assumptions that are somewhat ad hoc. In order to achieve acceptable fits to the observations, Reimers and Che-Bohnenstengel (1986) needed to include two acceleration zones for the wind: an inner flow that increased monotonically, plus an abrupt escalation to the terminal velocity. Schröder and Reimers (1989) found that an additional decrease in density coupled with the depletion of ground-state population of ions yielded an improved fit, though Fe II line profiles could still not be modelled satisfactory. Similar conclusions about the chromospheric density distribution, turbulence velocity and excitation temperatures were reached by Eaton and Shaw (2007). However, as they remarked, 22 Vul has probably had a different evolutionary history compared to the classical systems, perhaps involving mass transfer and enhanced magnetic-field activity; they cited as evidence its circular orbit, small component separation, and possible tidal locking. All three points warrant some comment. The component separation is sufficiently small to induce ellipticity effects in the light curve (Eaton et al. 1994), and the circularization of the orbit suggests that the system has undergone mass transfer. In their study (which also gave proof of the circularity of the orbit), Griffin et al. (1993) calculated that the rotation of the primary seems to be a little faster than its pseudosynchronous rate, but pointed out that extracting the rotational velocity from the *Coravel* 'dip' had probably underestimated the contribution from turbulence. That turbulence is now known to be higher for 22 Vul than the standard value which had been applied, and in fact the derived rotation rate would be exactly the synchronous value when a more appropriate value for turbulence is used. Another indication of mass transfer came from Schröder et al. (1997), who were unable to find evolutionary tracks that matched the G star and B star at the same time, the primary star being too luminous and too blue.

Complementing the *IUE* spectra described above, observations with *FUSE* gave a clear demonstration of the inhomogeneous outer atmosphere of this star, with

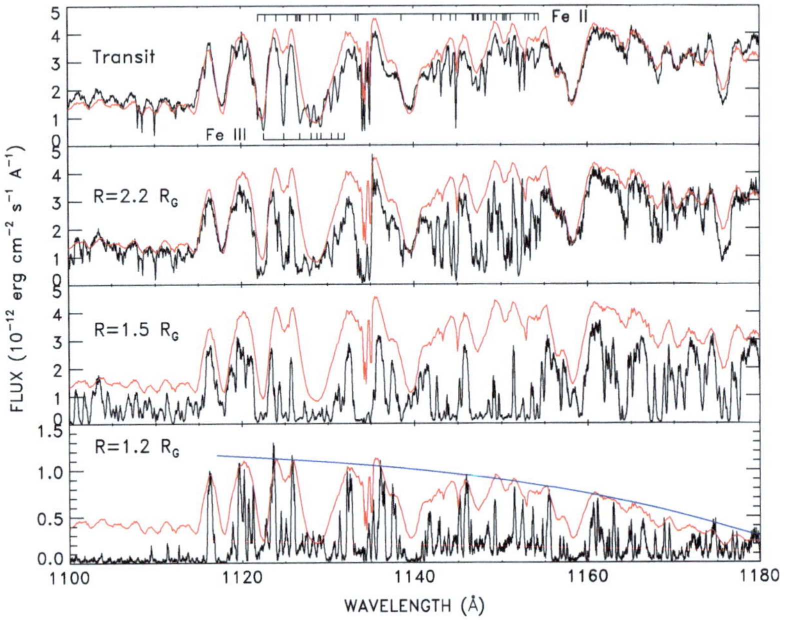

Fig. 2.24 Spectra of 22 Vul (*black*) compared to a scaled spectrum of a standard B9 V star, HD 32039 (*red*). The *top panel* shows that 22 Vul still exhibits Fe wind absorption features far from eclipse, when the B star is transiting the G star. The other panels are ingress observations, with the hot companion at various projected heights from the centre of the primary. Note the change in scale for the *bottom panel*. The reference spectrum has been scaled by the absorption expected from the Lyα wing in an atmosphere with a hydrogen column mass of 4×10^{22} cm^{-2} (*blue/smooth*)

components at distinctly different temperatures, or at least different degrees of ionization. Two weeks before total eclipse, the spectra revealed O VI in absorption at heights of 2.5 and 2.0 R_G (Ake and Robinson 2005). It was surprising to find material so hot (3×10^5 K) at that distance from the supergiant. When the continuum from the B star was subtracted, O VI λ 1032 Å was found to be in emission, with the absorption forming a centrally-placed reversal that matched spectra taken during totality. On the other hand, the C III line at λ 977 Å in eclipse showed evidence of a dense wind with a terminal velocity of -100 km s^{-1}. As in other coronal and hybrid supergiants, the O VI profiles in 22 Vul were symmetrical, in keeping with an origin in the supergiant's chromosphere rather than the shock front seen in C IV and Si IV, although the central reversal was found to be highly variable and unlike that seen in other late-type stars.

The out-of-eclipse spectrum and the changes seen during ingress in the wavelength region λ 1100–1180 Å are illustrated in Fig. 2.24. Fe II and Fe III features arising from the wind were visible at all phases, indicating that the secondary is always embedded in the wind of the supergiant. During ingress Fe II lines strengthen down to a distance of 1.7 R_G, when they start to saturate. The region from λ 1103 to 1120 Å is dominated by C I lines, which are not present at 2.5 R_G but strengthen

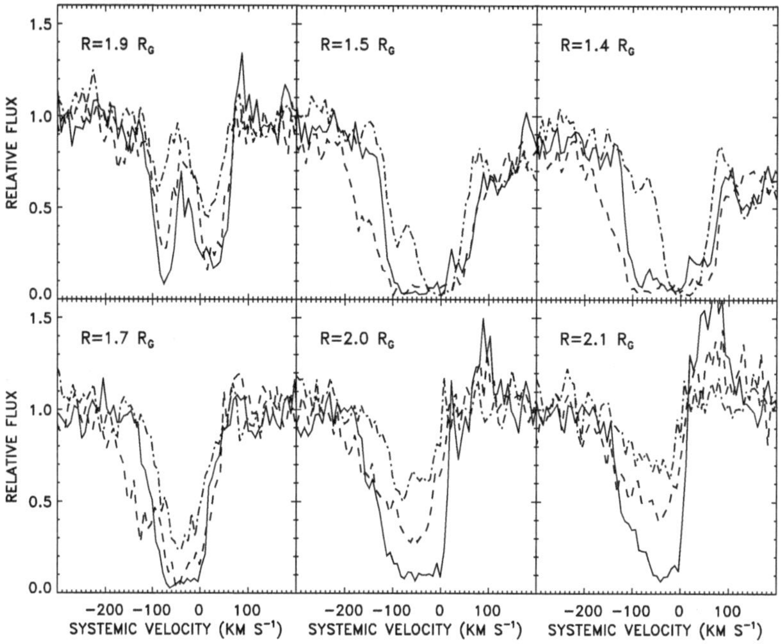

Fig. 2.25 Series of line profiles during ingress (*upper*) and egress (*lower*) phases of 22 Vul. 0-eV lines are represented by Ar I λ 1048.2 Å (*solid line*). Two excited Fe II lines with different log gf show opacity effects: λ 1071.6 (log $gf = -0.80$ *dashed line*) and 1089.6 Å (log $gf = -1.25$, *dotted/dashed line*). Multiple components appeared during ingress, but egress exhibited wind troughs typical of a smooth outflow. Red-shifted P Cygni emission in Ar I arose from resonance scattering in the wind

rapidly towards a maximum near 1.5 R_G, where C I opacity becomes dominant. The continuum absorption increases slowly at first, and then at a height of ~1.5 R_G it does so more dramatically, in harmony with an order-of-magnitude increase in the hydrogen mass deduced from the flux suppression in the wing of Lyα.

During ingress some absorption features developed double or multiple components and retained that structure over a range of heights (Fig. 2.25, upper panels). Different species exhibited their own characteristic patterns of enhancements and velocities which may differ from eclipse to eclipse, in a manner that is somewhat reminiscent of the changing profiles of the Ca II K line in 32 Cyg (Fig. 2.10). In contrast, during egress (Fig. 2.25, lower panels) the lines did not have the structure seen at ingress but showed the broad troughs and sharp redward cutoffs (depending on optical depth) that are more typical of wind lines, and with the same terminal velocity (about −200 km s^{-1}) as was seen during ingress.

In most late-type stars the spectrum from λ 1120 to 1150 Å shows fluorescent Fe II lines pumped by Lyα (Harper et al. 2001). However, when 22 Vul was in total eclipse, alongside those emission lines from the G star appeared a new complex of lines from λ 1142 to 1155 Å (Fig. 2.26). A 2-D model of the wind scattering

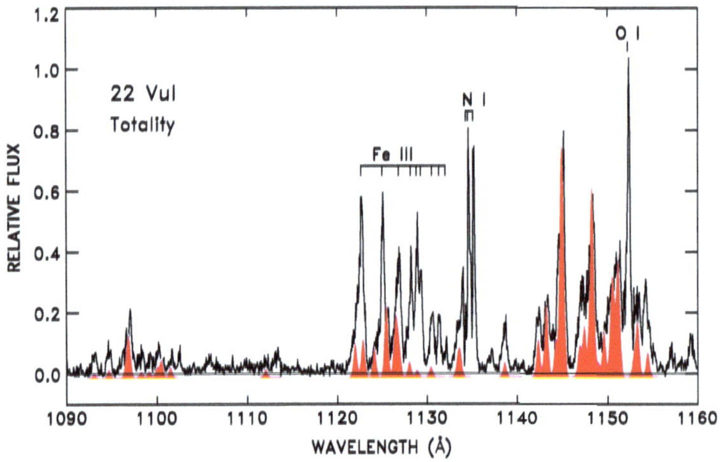

Fig. 2.26 Total-eclipse spectrum of 22 Vul (*black*) from *FUSE*, compared to a 2-D Fe II resonance scattering model (*filled/red*) of the B star in the expanding wind of the supergiant primary (Ake and Robinson 2005). The intense Fe II complex from λ 1142 to 1154 Å is a unique signature of wind scattering in 22 Vul. Other wind lines are noted

in Fe II by Ake and Robinson (2005) produced a reasonably accurate fit to the 22 Vul spectrum, aside from other emission lines that were known to be missing from the model. This region of the Fe II emission is therefore a unique marker of the resonance-line wind scattering by a hot star in a cool, expanding wind. That marker will make an important contribution in studies of the far UV spectrum of the enigmatic system ϵ Aur (Chap. 4, especially Sect. 4.2.5 and Fig. 4.4).

2.4.6 *HR 6902*

The hot companion of HR 6902 is of late-B type; it has only modest rotation (<15 km s^{-1}; Griffin 1986), and its spectrum includes enough sharp lines that the SB2 orbit can be determined with good precision. Schröder et al. (1997) were able to test their stellar-evolution models for this system with considerable exactitude. Figure 2.27 presents a scaled model of the geometry of the system.

2.4.6.1 Ground-Based Monitoring

Eclipses in HR 6902 were discovered in 1986. An analysis of its composite spectrum and SB2 orbit (Griffin 1986) had yielded $M \sin^3 i = 2.95\, M_{\odot}$ for the B8.5 V secondary, a value that was sufficiently close to standard tabulated masses as to suggest that $\sin i$ was close to 1.0. Photographic observations made at the DAO

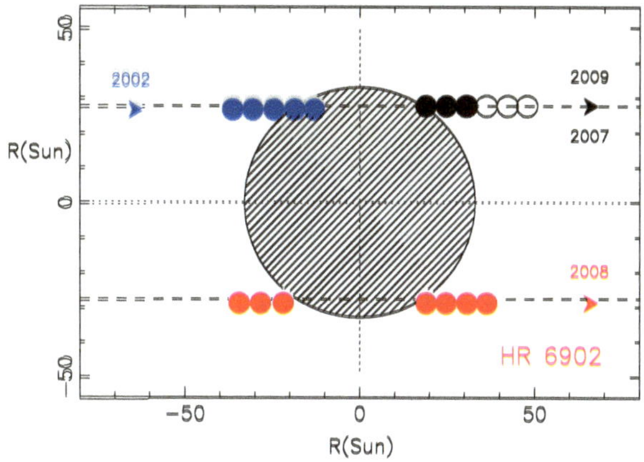

Fig. 2.27 Scaled geometrical model of HR 6902. $R_K = 33\ R_{\odot}$ (Griffin et al. 1995) was adopted. The trajectories of different eclipses have been located in opposite 'hemispheres' for clarity

encompassing the date of conjunction in 1985 proved that surmise to be correct (Griffin 1988), and a number of eclipses since then have been monitored from the ground (see Fig. 2.28).

The giant star in this system is both warmer and smaller than in the classical ζ Aur ones. Examination of partial-eclipse photometry and spectrophotometry shows that its chromosphere is less extensive even than the projected radius of the secondary, which means that it cannot be resolved by the light-probe method anything like as well as in the case of K-type supergiant primaries. To analyze optical spectra of HR 6902 obtained during the 1987 and 1989 eclipses, Schröder et al. (1996) employed a curve-of-growth method that had therefore to be modified (as described in Sect. 2.3) to take account of the real variation in optical depth across the projected image of the secondary during chromospheric-eclipse phases. The empirical density and excitation model which was derived provided some surprises. The number densities were very similar to those derived for ζ Aur at the same *relative* heights, i.e., when the heights were scaled by the radius of the relevant primary star, but only for heights less than about 2 $R_{\odot}$ in HR 6902, equivalent to about 9 $R_{\odot}$ in ζ Aur. Above that height, the column densities in HR 6902 decreased steeply and the ionization balance changed. But that did not signal the 'top' of the chromosphere: the turbulence then increased from 15 to 45 km s^{-1} and the density continued to fall, though much less steeply than closer to the photosphere, so the chromosphere's scale height was correspondingly larger there. Those properties suggest a separate component to the chromosphere that extends much further out than the more compact inner one. It is not a feature that has been detected in the K-supergiant chromospheres.

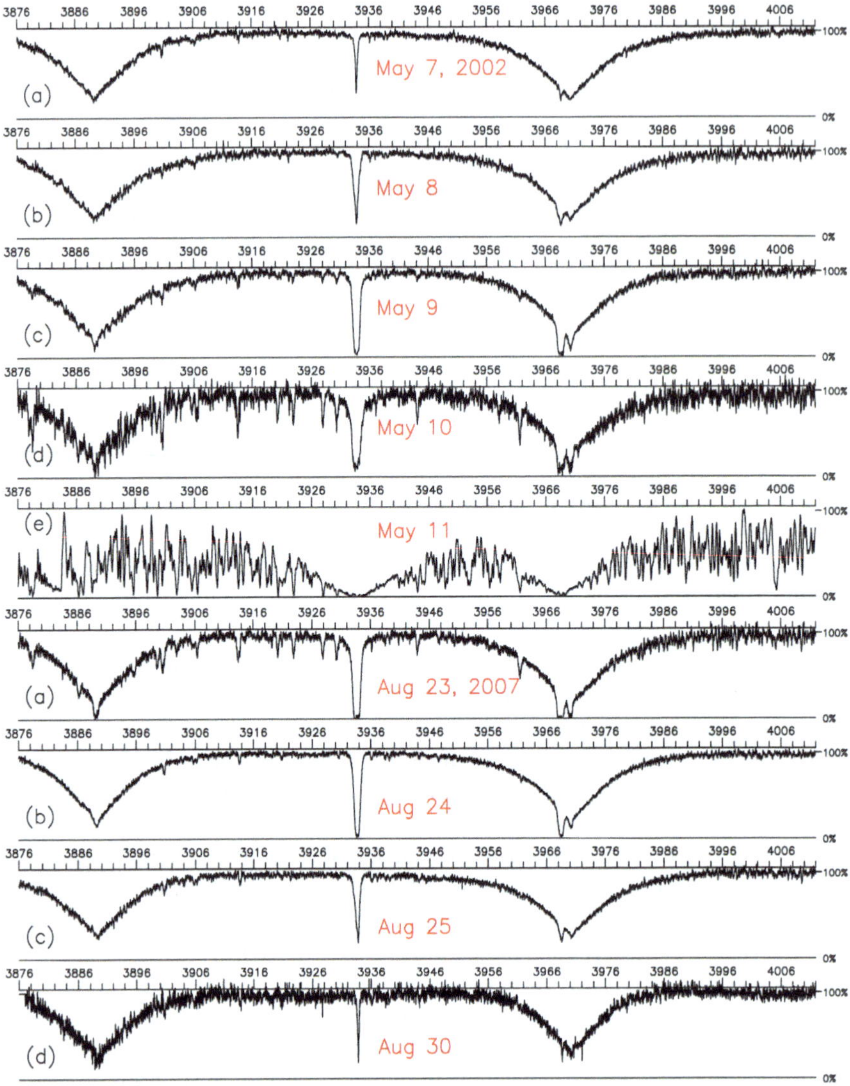

Fig. 2.28 Sequence of uncovered chromospheric spectra in HR 6902 during ingress in 2002 (*upper*) and egress in 2007 (*lower*) (Data source: CCD spectra obtained at the DAO by Griffin)

2.4.6.2 UV Spectroscopy from Space

IUE spectra of HR 6902 taken in 1988 spanned the eclipse of late August; four spectra were taken at the shorter wavelength region and therefore include the Si IV and C IV lines, which are widely regarded as diagnostics of a corona. Schröder (1990) divided the spectra taken during chromospheric-eclipse phases by ones taken far from eclipse, and derived column densities for ionized Mg, V, Fe and Si IV. The density of Si IV was found to decrease only slowly, indicating a large scale height with a possible physical extent to about 1 R_G, and an electron temperature of ~60,000–70,000 K, which was far in excess of that pertaining in the chromospheres of the cool supergiants. The Si IV absorption therefore represented an extended plasma at a temperature that was much more characteristic of the transition region than of the chromosphere.

A more extensive analysis by Reimers et al. (1990a) and reanalysis by Kirsch et al. (2001) of five *IUE* eclipse spectra taken in 1987 and 1988 found that emission fluxes in both Si IV and C IV were an order of magnitude greater in this giant than in other so called 'coronal' giants of similar spectral types. Since the primary itself could not be the origin of that excess, they divined that it represented evidence of a hot wind. To model that wind through Si IV, C IV, C II and Fe III lines, they employed Hempe's (1982) Sobolev approximation, deriving for the wind a mean electron temperature of ~90,000 K. They concluded that the excess strengths of the high-temperature emission lines seen during totality were caused by resonance-line scattering of B-star photons in that hot wind. Rapid expansion by the wind of some 30 km s^{-1} contributed to line broadening.

In contrast to all the systems discussed earlier, the chromosphere of HR 6902 is not extended; both the optical and the *IUE* spectra showed that it extends no further than ~0.1 R_G, whereupon it gives over to the higher-temperature plasma phase. Even at an effective height of 0.06 R_G the electron temperature which was derived is already 58,000 K. The greater surprise was therefore that those two distinct and very different structures could co-exist in the outer atmosphere of the star. A subsequent analysis by Reimers et al. (1990b) of *IUE* spectra of the 1989 eclipse added to the picture by revealing a complex ionization structure, together with densities and mass-loss rates that were greater than in a scaled solar-type model by several orders of magnitude. Without more spectra of suitable cadence it was not possible to define the temperature and wind-velocity gradients across the whole of that transition region, or therefore to determine the extent to which the properties of that outer region actually vary with time, though there was already evidence of changes between the two eclipses.

However, HR 6902 is not unique in these respects. As the following section outlines, a similar dual structure was found in HR 2554.

2.4.7 HR 2554

As with 22 Vul, the SB1 star HR 2554 (G7 II + A1 V) was not recognized as a composite-spectrum binary until *IUE* spectra revealed the presence of a hot companion; later spectra then recorded changes in the companion's flux that had to signify an eclipse (Ake and Parsons 1987). Eclipses are only partial, and chromospheric absorption lines can be seen for only 1 day before and after an eclipse (Ake and Parsons 1988). In that respect it resembles τ Per (Sect. 2.4.8), though

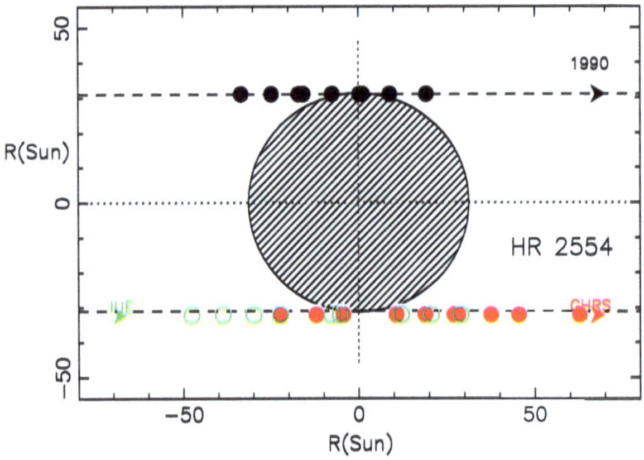

Fig. 2.29 Scaled geometry of an eclipse in HR 2554. A value for R_K of 31.3 $R_\odot$ was adopted from Brown et al. (2001). The trajectories of the two eclipses have been located in opposite 'hemispheres' for clarity. The upper set corresponds to ground-based observations made at ESO during the 1990 January eclipse. The lower one represents space-UV observations made with *IUE* at phases close to the 1987 eclipse (*open symbols*), and with *HST* during the 1994 and 1995 eclipses (*filled symbols*)

opportunities to observe its eclipses are relatively plentiful as it eclipses twice in most years ($P = 195.258 \pm 0.003$ days; Brown et al. 2001). Figure 2.29 depicts the scaled geometry of the system.

2.4.7.1 Ground-Based Monitoring

Optical (ground-based) spectroscopy of HR 2554 is not plentiful because the star is a southern object—it is presently the only known ζ Aur system in the southern sky. Eclipse photometry and an analysis by Schröder and Hünsch (1992) of optical spectra observed during the eclipse of 1990 January showed that an eclipse has a duration of 3.2 days, with a maximum immersion of ∼70% of the B-star's disk, and that the system's brightness falls in V, B and U by 0.06, 0.08 and 0.16 mag., respectively. From the colours deduced for the cool primary component they proposed for it a spectral type of G7 II.

As is the case for the other four systems with non-supergiant G-type primaries, the chromosphere of this system is less extended than the projected disk of the secondary (Schröder and Hünsch 1992). The lack of total eclipse introduced the unwanted complication that the spectrum of the giant could not be observed alone, and thus required the spectrum of a standard star to be selected in order to separate the component spectra of the binary (Fig. 2.30).

The chromospheric spectra that were obtained offered little time-resolved information about ions other than Ca II, or (therefore) about the degree of ionization in the chromosphere. Although column densities were derived for Ca II and Fe I,

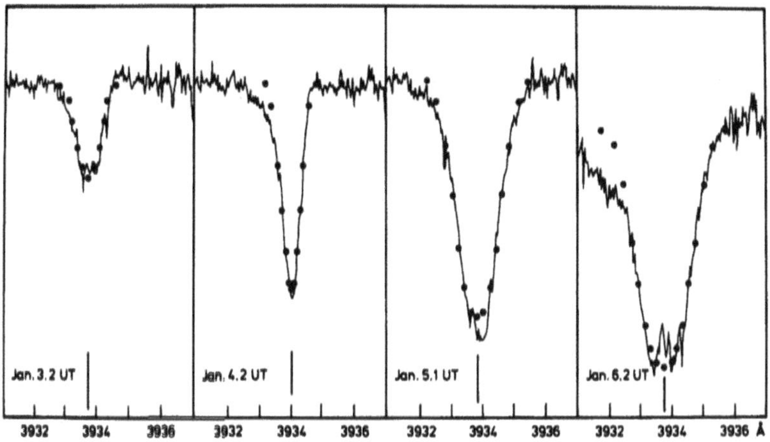

Fig. 2.30 Sequence of *K*-line profiles in HR 2554, extracted from observations made during eclipse ingress in January 1990. Spectra of the primary component were removed by a subtraction procedure (*Reproduced, with kind permission, from Schröder and Hünsch (1992)*)

they could not be converted into total particle densities. Nevertheless, the column densities that were derived for Ca II were similar to those found in HD 6902 (Sect. 2.4.6.1). Schröder and Hünsch also deduced that the chromosphere extended no more than about $\frac{1}{3}$ R_G, so the structure of the inner chromosphere of HR 2554 is not very different from that of HR 6902.

2.4.7.2 UV Spectroscopy from Space

High-resolution *IUE* spectra of HR 2554 (Schröder and Hünsch 1994) showed the same kind of evidence of material at transition-region temperatures (around 10^5 K) as had been found for HR 6902 (Sect. 2.4.6.2), though with intriguing differences. In HR 2554, column densities for Fe II proved to be similar to those derived for HR 6902 (Schröder 1990), but while the Si IV absorption was also similar, that for C IV was 4 or more times stronger. Even more intriguing was the discovery, by the X-ray space telescope *ROSAT*, that HR 2554 is an X-ray source, and possesses a coronal-like plasma at a temperature of $1–10 \times 10^6$ K in co-existence with material at least 10 times *cooler*. This evidence for hot outer regions in both HR 2554 and HR 6902 marks a significant departure from the extended but cool chromospheres and winds which encircle the more evolved supergiants, even though the *inner* layers of the chromospheres of these warmer G-type giants might be modelled as scaled-down versions of the cooler-star counterparts.

The analysis by Brown et al. (2001) of *HST* spectra of two later eclipses of HR 2554, plus eclipse photometry, yielded a precise determination of the parameters of the orbit and of the individual stars. They deduced masses of $M_G = 3.1\,M_\odot$, $M_A = 2.0\,M_\odot$, with uncertainties of the order of 5%, but also found that, when fitting theoretical stellar-evolution tracks to the pair in the H–R diagram, the G star appeared to be over-luminous and too blue for its mass. The inference was that mass transfer had taken place; if so, the structure of the G star may not be typical of that of an evolved single star.

2.4.8 *τ Persei*

The composite-spectrum binary τ Per has a period of 4.15 years. The possibility of eclipses was predicted by McAlister (1981) on the basis of speckle interferometry, and was confirmed photometrically by Ake et al. (1986), though the *IUE* spectra that were obtained led to an inconclusive decision about the existence of chromospheric features. That chromospheric absorption *did* occur was revealed in optical spectroscopy carried out at Lick Observatory in 1989 (Griffin et al. 1992). As Fig. 2.31 shows, eclipses are only partial, thus resembling those of HR 2554.

The mass ratio of 1.3 ± 0.1 given by the SB2 orbit (Griffin et al. 1992) is difficult to determine very precisely owing to the A-star's high rotational velocity of $\sim$250 km s^{-1}, but it was confirmed by rigorous matching of the positions of the components in the H–R diagram to theoretical evolutionary tracks (Schröder et al. 1997). Interestingly, both Griffin et al. and Schröder et al. concluded that the system has a metallicity that is a little lower than solar.

2.4.8.1 Ground-Based Monitoring

Eclipses in the τ Per system have an overall duration in V of barely three nights; the maximum obscuration of the secondary was calculated to be $\sim$80%. For a number of cycles subsequent to 1989, observations of eclipse phases were hampered

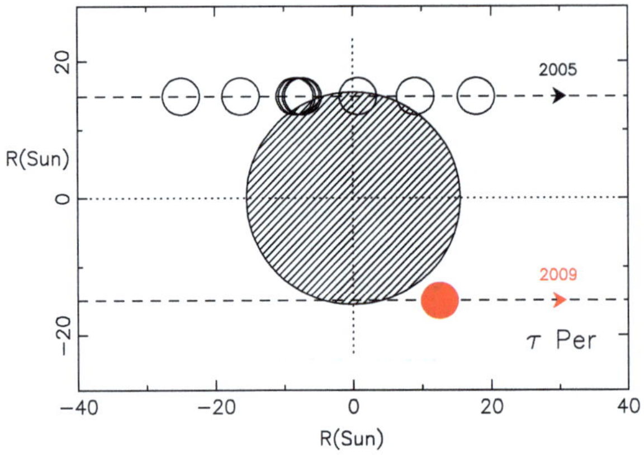

Fig. 2.31 Scaled geometry of an eclipse in τ Per. $R_K = 15.5\,R_\odot$ (Griffin et al. 1992) was adopted. The trajectories of the two eclipses have been located in opposite 'hemispheres' for clarity

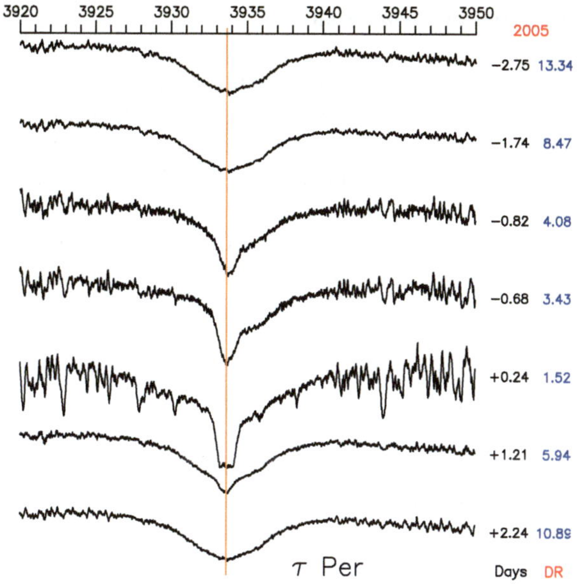

Fig. 2.32 Uncovered sequence of chromospheric spectra in τ Per during partial eclipse in 2005. 'Days' and DR are as for Fig. 2.3. The *vertical line* is merely to guide the eye (Data source: CCD spectra obtained at the DAO by Griffin)

by the slow migration of the dates into the season when the star cannot be reached, so the amount of data gathered has been rather limited.

An unbroken sequence of CCD spectra was obtained at the DAO during the eclipse of 2005 August; Fig. 2.32 shows the extracted secondary spectra from that sequence. Because eclipses are not total, a surrogate primary spectrum was required, and in fact that of the primary of γ Per proved to be a very close match. Conjunction was expected to occur on 2005 Aug. 23.00 UT, but because the star could not be observed until ~9 h UT the spectrum was recorded a little later than the time of central eclipse. The asymmetry which is apparent in the sequence in Fig. 2.32 can be fully explained by that 0.4-day offset; it was actually instrumental in refining the mid-eclipse timings of this fairly long-period system.

In Fig. 2.32, DR are the calculated distances of the *centre* of the secondary star, in $R_{\odot}$, from the limb of the primary. They should help to refine the determination of the actual size of the secondary, but they in turn depend on how the other free parameters (the relative tangential velocity of the component stars, the mass ratio, and the latitude of the eclipse) were also derived.

It was found that the chromospheric densities derived from the K-line profiles could not be matched by a uniform density distribution, suggesting instead a combination of asymmetry and inhomogeneities, or clumping, much as had been deduced for other ζ Aur chromospheres.

2.4.8.2 UV Spectroscopy from Space

Examination of the few available high-resolution *IUE* spectra led to the conclusion that τ Per has a solar-like corona surrounding a thin, clumpy chromosphere with a spicule-like structure at temperatures of some 2×10^4 K. An unusually small chromospheric scale height of less than 0.5 $R_\odot$ (0.035 R_G) was derived from Ca II (Schröder 1990), and the fact that the amount of Ca II remaining in the ground state was much smaller than that derived for other G-type ζ Aur systems led to the conclusion that chromospheric Ca II can become ionized to Ca III at quite modest temperatures. Because an eclipse in τ Per is rapid it is not easy to obtain observations at the correct timing and with adequate cadence. However, UV spectrophotometry from assorted *IUE* exposures obtained during or near the eclipse of 1984 November produced the puzzling result that the fluxes during eclipse were not as small in the UV as might be expected (Ake et al. 1986); the eclipse depth, measured as 0.35 mag. at λ 3200 Å, fell to 1.15 mag. at λ 2500 Å but remained at that level to λ 1600 Å and then rose to 0.25 mag. at λ 1250 Å. Unfortunately there were no *FUSE* observations of τ Per, and it remains unclear as to whether the system contains a third star or whether another cause was responsible for the observed effect.

2.4.9 *γ Persei*

The discovery that the 2.9-mag. system γ Per underwent total eclipses was not made until 1990 (Griffin et al. 1991; Schröder and Griffin 1991). That event elevated γ Per to the ranks of the famous, as the second-brightest visibly eclipsing system in the north, outshone only by β Per (Algol). Eclipses have depths of 0.28, 0.54 and 0.88 mag. in V, B and U, respectively (Griffin et al. 1994), and it may seem astonishing that such a pronounced event in such a bright star should have escaped notice for so long. But although an eclipse lasts for 9 days, failure to notice it could partly be attributed to the long period of 14.6 years, and partly to the fact that the formal uncertainty in the period, combined with a high orbital eccentricity of 0.78, had led to an ambiguity of many weeks in the date of conjunction. As described by Griffin et al. (1991), the matter was only eventually resolved when sufficient new RVs, measured with the high precision of the Cambridge RV spectrometer, enabled those uncertainties to be reduced.

Photometry obtained during the 1990 eclipse refined the date of eclipse to a precision of 0.01 day, and yielded radii of 24 $R_\odot$ for the giant, 3.2 $R_\odot$ for the dwarf, and an estimate of 34° for the eclipse latitude. Photographic spectra obtained during morning twilight time with the Palomar 200-in. telescope confirmed the eclipse, but limitations of telescope time and season did not offer much opportunity to maximize the S/N through repeated exposures. The subsequent eclipse in 2005 April, which was at a somewhat more favourable season, was monitored by CCD spectroscopy at the DAO (Fig. 2.33).

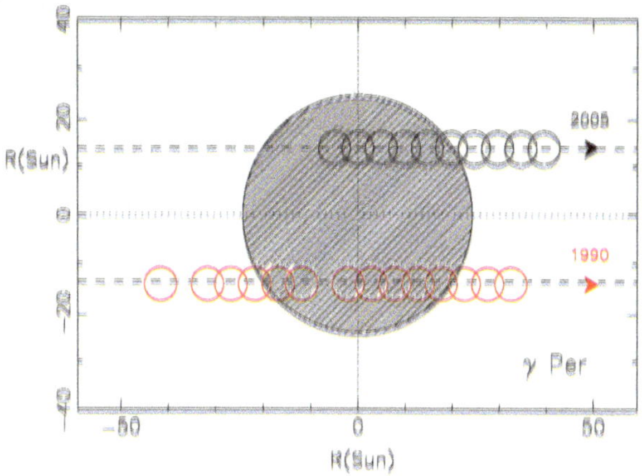

Fig. 2.33 Scaled geometry of an eclipse in γ Per. $R_K = 24 R_\odot$ was adopted from Griffin (2007). The trajectories for the two eclipses that have been monitored to date are drawn in opposite 'hemispheres' for clarity. The 1990 observations, which were photographic, constituted the 'discovery' spectra. A hint of chromospheric absorption was detected in the higher-S/N CCD spectra of 2005, and corresponded to a single phase when the secondary's disk was half uncovered (Fig. 2.34)

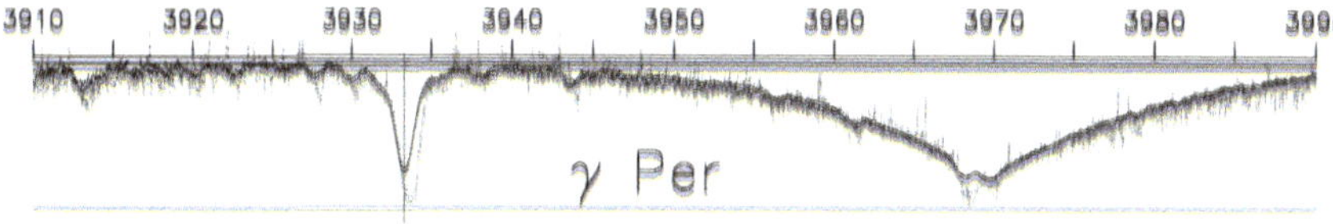

Fig. 2.34 Spectra of the early-A dwarf secondary of γ Per during egress from total eclipse in 2005. The *dotted profile*, recovered from the phase when the secondary was half-way out of eclipse, is our only evidence to date of any chromospheric absorption. The *continuous line* is a superposition of the spectra of the secondary on two subsequent nights. The *vertical line* marks the calculated position of the secondary spectrum observed (as here) in the rest-frame of the primary component. The substantial red-shift of the chromospheric *K* line is a consequence of the fact that the two stars pass one another obliquely rather than tangentially to the line of sight (Data source: CCD spectra obtained at the DAO by Griffin)

Unravelling the composite spectra has confirmed that the primary is a late-G bright giant and the secondary an early-A dwarf, and that they have a mass ratio of 1.50 ± 0.02. Analyses of the spectrophotometry, supported by the H–R diagram positions of the component stars, additionally suggests that both stars have raised luminosities. Griffin (2007) proposed spectral types near G8 II–III and A1 IV.

Figure 2.34 compares spectra of the secondary component, extracted from observations recorded during egress. The dotted line corresponds to the phase when the secondary was just half-emerged; the solid line is an average of the secondary spectra as extracted on the two subsequent nights. The transition from total eclipse to chromospheric eclipse was rapid, as was the subsequent transition to the end of eclipse. The vertical line combines the calculated orbital velocity of the secondary star ($-14.8\,\mathrm{km\,s^{-1}}$) with that of the primary ($+14.0\,\mathrm{km\,s^{-1}}$), in whose rest-frame it has been extracted, so the RV of the hot component was $-28.8\,\mathrm{km\,s^{-1}}$ at the time of that observation. Since the region of chromosphere being sampled at egress was at the approaching limb of the primary, the rotation of the primary would have imparted a blue-shift to the chromospheric line if the chromosphere shared the primary's rotation. In fact the measured RV of the chromospheric *K*-line in Fig. 2.34 was close to $+29\,\mathrm{km\,s^{-1}}$ relative to the secondary, implying that the primary had near-zero rotation. According to analysis of the *Coravel* observations the rotational velocity of the primary is only some 3–5 $\mathrm{km\,s^{-1}}$, and a velocity shift of that amount was probably too small to measure very precisely in that one egress spectrum.

That meagre displaced absorption in the *K* line is all the observational evidence so far to hand of the chromosphere of this giant. Its apparent thinness or near-absence could mean that the chromosphere is sufficiently warm that Ca II is mostly ionized to Ca III, as was deduced for τ Per (Sect. 2.4.8.2). The same may also be true of HR 6902 (Sect. 2.4.6) and HR 2554 (Sect. 2.4.7); all four G-type primaries have substantially higher surface gravity than a K-type supergiant. In γ Per there was also no hint of absorption in the cores of the Balmer lines. Detection of the chromosphere in this system is further compounded by the relatively large size of the secondary star, which dilutes the chromospheric flux and degrades the spatial resolution compared to the systems with proportionately larger primary stars.

Although offering little or no evidence of a cool chromosphere, the spectra combined with the measured flux ratios of the two component stars have contributed usefully towards firming up the timings of second and third contacts.

2.4.10 HD 223971

HD 223971 (G7 III + F2 IIIm) is a composite-spectrum binary in a circular orbit with a period of 50 days. An SB2 solution for the primary yielded the high mass function of 0.35 $M_\odot$, suggesting that eclipses were possible; alerting a network of amateur photometric observers led to the discovery of an eclipse in August 1990 (Griffin et al. 1991). However, since totality has a total duration of about 19 h, it can be difficult to monitor it effectively if the phases do not match the hours of darkness.

Since it is an eclipsing composite-spectrum binary, HD 223971 should in principle should be classed as a ζ Aur system. In practice, observational evidence—if there is any to be had—of its chromosphere is difficult to come by, for the following reasons:

1. The relatively short period means that the component stars pass one another more quickly than in any of the other ζ Aur systems, so very short exposures would be needed to capture chromospheric spectra during the brief intervals when they might be visible.

2. A model of the system indicates that the cool primary and warm secondary have radii of 6.9 and 2.9 $R_{\odot}$, respectively, so the spatial resolution of any chromospheric material will be much degraded.
3. The secondary star, an early F-type, has a fairly rich spectrum of its own. Its lines will tend to mask any signatures from the giant's chromosphere, particularly since the orbit is circular so the only observable velocity displacement between either star and any chromospheric lines will be that attributable to the rotation of the chromosphere. Experience from the case of γ Per suggests that such evidence will at best be slender.
4. Since the star has $V = 6.6$ mag. and the secondary shows rather little violet flux, obtaining adequate S/N at adequately high dispersion at H and K wavelengths during the short exposures that are now seen as necessary is challenging. No certain signatures of the chromosphere were revealed in the K line from a sequence of spectra observed through ingress in 1991; the observed strengthening of the late-type lines that characterize a chromospheric absorption spectrum could be fully attributed to an increasing dominance by the giant component as totality approached.
5. The cores of the H and K lines in the giant itself contain emission components which tend to hamper the detection of any extra absorption from the chromosphere.

The scaled model of the system in Fig. 2.35 illustrates the problem. Observations in 2001 were made on adjacent nights; the two in 2003 were separated by 48 h. The absence of any obvious features of chromospheric origin in the spectrum during egress in 2001 places a strong constraint on that model. HD 223971 may therefore

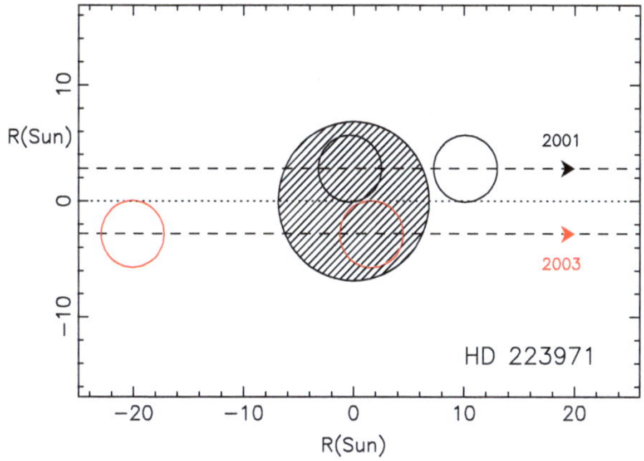

Fig. 2.35 Geometry of an eclipse in HD 223971, drawn to scale. R_K of 6.8 $R_{\odot}$, derived from unpublished eclipse photometry, was adopted. The trajectories of the two eclipses have been located in opposite 'hemispheres' for clarity. Eclipses are swift; the two observations made in 2001 were only 1 day apart

represent the limit of what chromospheric evidence can be detected in the optical region by the light-probe method.

However, the steady decrease in visible evidence of a chromosphere as instanced by the sequence of systems described earlier must be due to the method of detection rather than to any actual lack of a chromosphere. Evidence from the Sun is proof enough that G-type stars possess chromospheres, but if the plasma is doubly ionized it will not be detected in features like Ca II or Ti II in the optical region. Space-UV spectra of HR 6902 and HR 2554 have shown that these hotter giants (which are also less compact than the Sun) are surrounded by plasma at transition-region or near-coronal temperatures (Sects. 2.4.6.2 and 2.4.7.2), and that the region of more 'traditional' chromospheric absorption, detected in lines like Ca II, has shrunk to a thin layer that is substantially smaller than the projected diameter of the secondary star. This increasing lack of optical evidence may therefore be as much a limitation in the physics as in spatial and temporal resolution.

2.4.11 ϵ Aur

It must be emphasized that ϵ Aur is *not* a ζ Aur system. However, as described below, features seen during the ingress and egress phases of its principal eclipse do offer a link—if a misleading one—between the ζ Aur systems and the full review of this star given in Chap. 4. Spectroscopy of the Ca II *H* and *K* region (Fig. 2.36) provides a helpful introduction to some of the challenges presented by this object.

While it is immediately clear from Fig. 2.36a that the primary star in ϵ Aur has a considerably earlier spectral type than any of the ζ Aur primaries described above, the appearances of what were often called 'shell lines' during eclipse ingress and egress (Fig. 2.36b) do bear a curious resemblance to the growth of chromospheric absorption components which blend with their photospheric counterparts during phases of chromospheric eclipse in a ζ Aur binary. The two spectra in Fig. 2.36b correspond to early ingress (drawn thick) and to late egress (drawn thin). The only

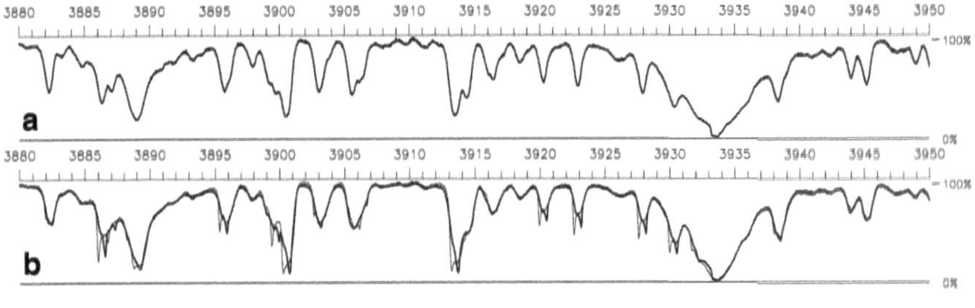

Fig. 2.36 The spectrum of ϵ Aur in the near-UV. *Upper*: the spectrum far from eclipse. *Lower*: Two superimposed spectra observed during early ingress (*thick line*) and late egress (*thin line*). The 'shell' lines in the lower plot are considerably sharper than the profiles of the same lines in the F-supergiant primary, and are mutually displaced in velocity by about 65 km s^{-1}

major differences between them are the RV displacements of the sharp spikes that are visible in the cores of ground-state and low-excitation lines of Fe I. They are also in the same sense as is observed for chromospheric absorption lines. In a ζ Aur binary the rotation of the primary and at least of its low chromosphere usually causes the chromospheric lines to be red-shifted during ingress and blue-shifted during egress, and in ϵ Aur the shell lines, which are formed in the outer parts of the rotating opaque disk as it transits across the F supergiant, are also correspondingly displaced.

Recent progress in understanding (or tackling) the system of ϵ Aur is presented in Chap. 4 (page 107).

2.5 Modelling Stellar Chromospheres

2.5.1 Densities in the Lower and Middle Chromosphere

The extraction of an average density distribution from pure absorption-line spectra appears to give valid empirical models for intermediate chromospheric heights—i.e., above the dense, emitting base but below the wind-acceleration region where the velocity field becomes comparable to the speed of sound. It makes the minimal assumption that the chromosphere has only radial variation, though it must clearly be limited by that simplification and it does not address the inhomogeneous and dynamic nature of the chromosphere or wind. As a matter of fact, the kinds of severe inhomogeneities that would invert the density gradient or cause lines to split are not typical for most ζ Aur systems, and have been observed in only a minority of cases (e.g. in 32 Cyg: Schröder 1983). The divergence of the wind velocity field in the line of sight, and other large-scale contributors to the line-broadening velocity, while not strictly stochastic in nature, still have sufficiently small optical-depth scales to mimic the reduced line saturation as would be caused by supersonic turbulence. Thus, while line-broadening velocity as such has no physical meaning in describing saturation in a curve of growth or a line profile, the resulting column densities may nevertheless be reasonably accurate. It is also safe to neglect residual emission, as it would potentially affect only very strongly saturated line cores and not the optically thin lines that are preferentially used to determine column densities by the curve-of-growth method.

However, a serious limitation that emerged in the application of the simple attenuation theory was the need to include an apparent stochastic line-broadening velocity in the range 12–25 km s^{-1} (i.e., supersonic) at the same time as hydrodynamic turbulence, which was necessarily subsonic. Incorporating a diverging mean velocity field to describe regimes of different radial motions which, though optically thin, were also macroscopic in extent, was challenging for the analysis of the outer regions where the wind acceleration begins.

The results obtained from spectroscopy of the classical ζ Aur binaries and VV Cep at the DAO and at Mount Wilson prior to 1970 (e.g., by Christie and Wilson 1935; Wilson 1948; Wilson and Abt 1954; Wright 1959) provided basic physical parameters of the chromospheric gas, and are summarized here before the results derived from post-1970 efforts are reviewed. Numerous papers (e.g., Wright 1952, 1970; Wright and Lee 1959; Lee and Wright 1960; Popper 1961; Wright and Huffman 1968) also defined the physical properties of each binary system and its components. The research was based almost exclusively on lines of Fe I and the ground-state Ca II *H* and *K*, but with a smattering of lines from Fe II and a few other singly-ionized metals. It therefore focused on the phases when those features were most visible, and those corresponded to the lower (denser) 0.5 R_K or so of the chromosphere and inner wind. Analyses derived column densities along the line of sight, and deduced excitation temperatures plus estimates of turbulent velocities from the random motions of the gas in the chromosphere and wind.

Although each system proved to be unique in detail, some general properties about this region of a chromosphere could be offered:

- The density decreases with height,
- excitation temperatures appear to rise with height,
- turbulence velocities are substantially greater than in the photosphere, and reach 10–15 $\mathrm{km\,s^{-1}}$ in the outer layers,
- metals are predominantly singly ionized, but nevertheless
- neutral species such as Fe I are unexpectedly strong, so ionization is much less than expected for gas in the radiation field of a B star, and
- electron densities, of $\sim 10^{10}\,\mathrm{cm^{-3}}$, exceed what is expected from a smooth distribution, implying significant condensations in the chromosphere or wind. Similar clumping was deduced by Osterbrock and Flather (1959) for the Orion Nebula.

What most seriously impaired the quality of the early attempts based on *analogue* optical spectra was not so much the suitability of the method, but the inability to isolate the chromospheric absorption lines from close blending with the photospheric spectrum of the cool giant, which contained the same lines but in different relative strengths.

2.5.2 The Upper Chromosphere and Wind

Access to UV spectra from space inspired fresh attempts to recognize and model what could now be detected. The profiles of the *K* line and the strengths of low-excitation or easily-ionized species could yield uncontaminated column densities for the chromosphere on an assumption of pure absorption at heights where contamination by scattering could safely be neglected, but space-UV spectra revealed other important components to the picture—increased energy and turbulence of the upper regions, an expanding wind, and the environment of the hot star and its interaction with the wind—that required a refinement of the routine physics, and 3-D computing codes. The complementary nature of the challenges lent an exciting new dimension to the understanding of these stars. Despite the inadequacies of laboratory data for those wavelengths, major new results were achieved:

- The mass column density of H I was derived from direct observations of Lyα; similar analyses of the ISM by Savage and Jenkins (1972) and of the winds of M supergiants by Snow et al. (1987) and Buss and Snow (1988) paved the way. Eaton and Bell (1994, their fig. 2) demonstrated how the strength of Lyα increases over a mass column density range from 10^{-5} to 0.3 g cm^{-3}, and illustrates how the effect of scattering by H I can easily be measured in the wings of Lyα except when column densities are small enough to be dominated by the photospheric profile of the B star. Figure 2.12 shows examples of Lyα absorption in the spectrum of 31 Cyg, and the mass column densities that had been determined.
- Column densities derived for H, Fe, Si and other elements were self-consistent. Hydrogen was mostly neutral throughout the chromosphere or wind; metals were singly-ionized. Ionization was higher in the vicinity of the B star.
- Turbulence, measured as 10–15 km s^{-1} in the outer chromosphere, was fairly high for such stars.
- The ionization balance implied a higher electron density than seemed possible since most of the electrons came from the metals, although the electron densities were still lower than those inferred from optical spectra.
- Precise excitation temperatures were determined from measurements of Fe II multiplets at many different heights in the outer atmosphere. The temperature rose with height and attained values that accorded with predictions from semi-empirical models (see fig. 6 of Eaton and Bell 1994).
- Part of the considerable line broadening that was observed was due to differential wind expansion, as was demonstrated for 31 Cyg (Fig. 2.13). Eaton (1996) mapped a radial expansion in a wind onto the line of sight to demonstrate how it broadens the absorption line profiles; the contribution from actual random motion may be as little as half the measured broadening. Kirsch and Baade (1994) and Baade et al. (1996) arrived at a similar conclusion for ζ Aur.
- The density and velocity structure of the wind could be determined by modelling the strong emission lines that are scattered from the B-star's continuum by the cool supergiant's wind (and demonstrated in the case of ζ Aur by Baade et al. 1996).
- The degree of departure from LTE can be assessed by comparing Balmer-line strengths with Lyα to obtain a direct measure of H excitation, and thence an estimate of the departure from LTE at various levels in the chromosphere (Table 2 of Eaton and Bell 1994). The same check can be applied to Mg II, using the four lines near λ 2800 Å (loc. cit., Table 4). The result offered the potential to test semi-empirical models of the chromospheres, but does not appear to have been tried yet. The same departure coefficients could also be derived in the optical region by combining Balmer line strengths with inferred H I column densities.

- Intrinsically weak lines that are too weak to be scattered from the B star's spectrum were found to have about the same strengths in the ζ Aur spectra as in single K stars (Schröder et al. 1988; Eaton 1992). The ζ Aur binaries therefore constitute a useful means for defining the chromospheric structure of similar supergiants (but see Sect. 2.9).

2.6 Comparisons and Contrasts

The usefulness of comparing and contrasting supergiant and giant chromospheres will depend upon the credibility of the analyses, which in turn are dependent upon the resolution and cadence of the data and the degree of sophistication which the models achieve. As already evidenced in many different ways, a chromosphere is neither static nor homogeneous, so models based on individual eclipse mappings can only relate to the gas that happened to be in front of the hot star, as projected onto a given trajectory, at that one time. The nature and characteristic sizes of any clumpiness in a chromosphere constitute additional unknowns.

In one of the first studies of cool-star chromospheres based on space-UV spectrophotometry, Doherty and Jung (1975) used quasi UV photometry of 31 Cyg and 32 Cyg from the OAO–2 satellite (10-Å bands from λ 1430 to 4250 Å) to investigate the strengths of overlapping multiplets of Fe II. They found that absorption by Fe II could be followed to much greater heights than had been detected in ground-based spectra, and surmised that the chromospheres of those two stars are substantially more extensive than previously believed. They also found that while both chromospheres could be described by the same density profile, the density in the chromosphere of 32 Cyg seemed to be greater than that of 31 Cyg but they could not identify the reason unambiguously. However, the two stars do exhibit different characteristic line profiles during atmospheric eclipse, possibly because their eclipses sample quite different regions of the chromosphere. It may be relevant to recall the evidence of slow fluctuations in the activity of the chromospheric material seen in Fig. 2.17, and the noticeable decrease by 2012 in the number and energy of random clumps compared to what had appeared in 2009 and increasingly so in 2006 and 2003. The suggestion that the structure of a cool-supergiant chromosphere undergoes the kind of transition from small-scale spikiness to broader, smoother streamers that is seen periodically in the structure of the solar corona is a conjecture that will take many cycles of eclipses to prove or refute, but could be tackled through robotic acquisitions of ground-based spectra.

2.6.1 How Similar Are These Chromospheres to One Another?

In view of the differences between the cool-supergiant primaries of ζ Aur and 22 Vul in effective temperature (3960 and 4700 K, respectively), in physical size (22 Vul is only about $\frac{1}{3}$ of the size of ζ Aur) and in luminosity (22 Vul is nearly a magnitude fainter in M_V than ζ Aur), the densities in their chromospheres are surprisingly similar. In both cases a mean excitation temperature for the chromosphere of about 5500 K was found, a value that appears to be typical for the chromosphere of a cool star. As indicated in Sect. 2.4.1.1, the strengths and profiles of pure-absorption optical lines in ζ Aur yielded hydrogen particle densities, log n_H, of 11.7 cm^{-3} at a height of 3.3 $R_\odot$ from the photosphere, falling only slightly at the greater height of 8.8 $R_\odot$, so the scale height of that chromosphere is of the order of 4–6 $R_\odot$ (and is in fact a little smaller than its physical height). For 22 Vul, the model which Schröder et al. (1994) derived (again by analyzing pure-absorption optical spectra; see Sect. 2.6.1) was not substantially different, given the likely uncertainties: hydrogen particle densities of 11.3 cm^{-3} at a height of 2.8 $R_\odot$ from the photosphere, and a scale height of 2.8 $R_\odot$ between physical heights of 3–6 $R_\odot$. However, that similarity in densities ceased above a height of 9–10 $R_\odot$, where Fe I and Ca II revealed strong but unequal degrees of photo-ionization that increased outwards, presumably under the influence of the flux from the companion; the decrease in densities was less steep in 22 Vul, whose late-B companion has a temperature of 11,000 K, than in ζ Aur, whose mid-B companion is at ~15,000 K.

In contrast, the G-type giant in HR 2554 has a higher gravity than 22 Vul; it is also slightly warmer, and in the H–R diagram the position of its primary is further from the cool-wind region than is the supergiant in 22 Vul (Schröder and Hünsch 1992). The geometrical extent of the chromosphere of HR 2554, when defined by the presence of measureable Ca II absorption, is much smaller than that of 22 Vul, and its Ca II column densities are four orders of magnitude lower (see Sect. 2.4.7.2). Schröder and Hünsch found the geometry of its grazing eclipse helpful for resolving the lower chromosphere, but it presented a technical difficulty for isolating the chromospheric features by subtraction as there was no opportunity to observe the giant's spectrum alone.

As described in Sects. 2.4.7.1 and 2.4.7.2, the scale height of the chromosphere of HR 2554 (about 0.2 R_G) is still proportionately similar to those of 22 Vul and ζ Aur, though the lower column densities that were derived in even the deep partial-eclipse phases indicate a rather high degree of ionization of Ca II throughout the chromosphere. The early-A dwarf component of HR 2554 is much cooler than the mid-B dwarf in ζ Aur, so the similarity between its chromosphere and those of the cooler supergiants must be a property of the chromosphere itself.

2.6.2 How Different Are They?

There is an obvious separation between the two groups of ζ Aur systems identified in Sect. 2.1: those with cool (super)giant primaries, and those containing G-type giants or bright giants that are less luminous and extended. τ Per, HR 2554 and HR 6902—the three systems of the second group for which quite reliable analyses of the low chromosphere are available—stand apart from the former group, as might be expected from their H–R diagram positions vis-à-vis the Linsky and Haisch (1979) 'dividing line' (Sect. 2.1). They share certain resemblances, but are also distinguished by differences, such as their outermost temperature structure and the co-existence of plasmas at very different temperatures, which could only be recognized through space spectroscopy. With regard to the lower chromosphere, much that has been deduced for the G-type giant primaries is shared. Although τ Per has a surface gravity that is two or three times larger than the other two G-giants and is also the warmest of the three, its chromospheric characteristics are quite similar to those of HR 2554 but like a scaled-down version. It has small column densities that are visible only in a strong Ca II K line, and the observational evidence suggests much smaller scale heights, of the order of 0.1–0.6 $R_{\odot}$. As described in Sect. 2.4.8, the total extent of the chromosphere that can be detected in τ Per is less than 1.5 $R_{\odot}$ (Griffin et al. 1992). The inner region of HR 6902 is also similar to that of τ Per in most respects, though variability from one eclipse to another in both systems, presumably stemming from inhomogeneities or clumping of some form, requires multiple monitoring sequences in order to identify the baseline properties of each. HR 6902 bears an interesting resemblance to HR 2554 in that both show evidence of a plasma at transition-region temperatures of some 1–10 $\times$ 10^5 K alongside chromospheric material which is at only 10^4 K (Sect. 2.4.7.2). τ Per too shows evidence of a hot solar-like corona, but the rise that was recorded in the UV flux towards shorter wavelengths during eclipse needs to be understood before firm conclusions can be drawn concerning its outer atmosphere. The secondary star also presents an unresolved puzzle concerning its high rotation rate and reduced metallicity. Both of these unexplained features appear to be unique to the τ Per system.

There are other interesting differences among the ζ Aur primaries as a group. In ζ Aur the symmetrical splitting of the K line in egress, and in other absorption lines of once-ionized species that persisted to considerable radial heights, was curiously well-defined; the maximum separation recorded was $\sim$40 km s^{-1}, with only small variation between eclipses, and both of the split components maintained the same limits. Patterns of that nature have not been seen in the K lines of either 31 Cyg or 32 Cyg. On the other hand, in 22 Vul some space-UV Fe II lines have appeared doubled, though only during ingress; the same lines were single during egress.

As mentioned in Sect. 2.1, 22 Vul appears to straddle the interval between the cooler systems and the hotter ones, and that is true of its FUV spectra too. Though late B in type as opposed to the early- or mid-B types of the classical ζ Aur systems, the hot companion of 22 Vul yields a very adequate signal at those short wavelengths. HR 6902 does as well, though not as intensely as 22 Vul.

The near-absence of detectable chromospheric signatures in both γ Per and HD 223971 suggests that those systems represent more extreme scaling of the τ Per model; in those cases, only upper limits can be assigned to their chromospheric properties.

2.7 The Inhomogeneity of the Chromosphere

All the analyses of these stars have found larger electron densities than would be expected if the gas were evenly distributed in a medium that reached smoothly from the cool-star's photospheric surface to interstellar space. The inference is that the gas is clumped, perhaps dramatically so, but the result has always been somewhat controversial since semi-empirical model chromospheres for single stars have of necessity been homogeneous, yet have still appeared to explain adequately the emission-line spectra of such stars.

Clumping may take different forms. The gas could be confined to magnetic flux tubes as in the Sun, for example—and one interpretation of the splitting in the *K* line of ζ Aur (Sect. 2.4.1.1), and in particular its confinement in velocity space and its frequent non-occurrence, has already suggested the possibility of such structures at certain heights. Alternatively, the chromospheric gas could be distributed in bunches of blobs, as in traditional models of gaseous nebulæ. In view of the partial ionization that is found in those atmospheres, the unexpected strengths of the neutral species may merely reflect the existence of denser zones of neutral gas throughout the chromosphere and wind, perhaps compressed by packets of ionized material that are being driven by Alfvén waves (see Chap. 5, page 123). However, if the gas is clumped microscopically (as in gaseous nebulæ), there have to be instabilities that break it up into clumps, alongside forces that are keeping it clumped. The logical explanation is that those forces are the ones which extend the chromosphere and drive the wind. For 31 Cyg, Eaton (2008) constructed a semi-empirical model based on that consideration, and found that it predicted moderate clumping within factors from about 2 to 10. Nevertheless, while investigating a chromospheric model for ζ Aur, Harper (2010) found that *less* clumping is required to explain the intrinsic radio emission.

Inhomogeneity may also show up as global differences in the mass of the chromosphere during ingress and egress of individual eclipses, or in the eclipses of the same star on different occasions. In optical studies, differences have been recorded in the strengths of chromospheric absorption features between ingress and egress of the same eclipse and between different eclipses of the same star (Figs. 2.3–2.5). Differences of that nature were also summarized by Wilson and Abt (1954) and by Wright (1970, figs. 3, 4 and 9). In the UV, *HST* spectra of the 1993 eclipse of ζ Aur revealed very different amounts of gas in ingress compared to egress (Baade et al. 1996).

The line-splitting that is occasionally seen at high values of R_K in the *K* line of ζ Aur is a direct manifestation of inhomogeneity in the upper chromosphere, but it seems to be unique to that system. It is not a disconnected random event, but has a more controlled and repeatable nature. The observation that the chromospheric *K* line is stronger when line-splitting occurs, and that the splitting is symmetrical and almost constant in velocity, suggests a regular structure that is intermittent in space and consists of isolated concentrations of tightly confined ionized material that are encountered only between certain heights, not unlike solar flux tubes. Between the structures are large and relatively empty gaps. What is stochastic about the splitting is not its magnitude and direction but whether or not there is such a structure in front of the projected trajectory of the hot star. Why splitting is not observed, or is not so pronounced, in ζ Aur during ingress could be an asymmetry that is attributable to the influence of the B-star's flux on the upper chromosphere (Sect. 2.4.1.1); at the egress phase when line-splitting has been most prominent, the inter-star distance is only 83% of what it is at the corresponding ingress phase, so the effect of the B-star's flux will be about 45% greater then compared to ingress.

2.8 Mass Loss and Stellar Winds

2.8.1 Properties of the Winds

The evidence for winds and mass loss in these binary systems is incontrovertible. Ca II and UV spectra provide evidence of upwelling material at all phases of atmospheric eclipse. In both ζ Aur (Baade et al. 1996) and 31 Cyg (Eaton and Bell 1994; Eaton 2008) the observed velocity structure of the absorption features is consistent with material accelerating away slowly from the K supergiant. That slow acceleration has been verified through the detection of radio emission from the material in an ionized zone around the hot component of ζ Aur (Harper et al. 2005). Mass-loss rates determined from the strengths of the wind lines are of the order of a few $10^{-8}\,M_\odot$ year^{-1} in all three K-supergiant systems.

In the case of 31 Cyg the wind is isotropic to a first order, inasmuch as *IUE* detected it moving toward the hot component in lines of ions that were triply-ionized by the hot star (sect. 3.6 and fig. 11 of Eaton and Bell 1994). Taylor and Seaquist (1984) predicted the presence of over-ionized zones in the winds of cool-binary components that enveloped their hot companions, as also did Che et al. (1983) for the ζ Aur systems. That hydrogen is ionized in such a zone in 31 Cyg is shown by the velocity structure of O I both in absorption (fig. 10 of Eaton and Bell 1994) and in emission (fig. 7 of Bauer and Bennett 2014).

To model the wind lines of ζ Aur systems, Hempe (1982) developed an approximate 2-D transfer code that employed the local escape probabilities on a grid of wind points, according to the Sobolev approximation. In pioneering work that compared wind-line profiles in *IUE* spectra and their phase dependence upon such models, Che et al. (1983) derived the first mass-loss rates for the classical ζ Aur systems, finding values ranging from 0.6 to $2.8 \times 10^{-8}\,M_\odot$ year^{-1} and wind velocities from 40 to 80 km s^{-1}. They needed to use a steep wind-acceleration prescription, $v(r) = v_\infty(1 - R_*/r)^{0.5}$, plus a large microturbulence that was clearly much larger that it could be in reality, in order to achieve good agreement between model and observation, but subsequent investigations of the outer chromospheres of those K-type supergiants by Schröder (1985) showed that the wind acceleration is much steadier, and is better described with an exponent of $\sim$2.5 $\pm$ 0.5. The smaller ζ Aur systems which contain G giants provided evidence that the cool wind gradually disappears towards the 'wind dividing line' of Mullan and Stencel (1982). HR 6902 and HR 2554 showed geometrically extended absorption from C IV and Si IV lines, rather than from the Fe II and Si II wind resonance lines that are seen in the classical ζ Aur systems (Reimers et al. 1990b; Schröder and Hünsch 1992), suggesting that the thermally unstable conditions of the solar transition region could even describe the whole wind region of those G-giant stars.

The improved methodology of a code that used an exact radiative-transfer solution and incorporated a realistic wind-acceleration law (Baade et al. 1996) removed much of the previous requirement for a microturbulence that was too large to be physical. For ζ Aur the terminal wind velocity which was then derived was about 70 km s^{-1}, and the mass-loss rate about $5 \times 10^{-9}\ M_{\odot}\ \mathrm{year}^{-1}$.

A flow of wind towards the hot star can also be detected through its interaction with that star. Ahmad et al. (1983), Ahmad and Parsons (1985) and Ahmad (1986, 1989) observed accretion in a column of gas formed behind the bow shock of the wind in several of these systems, using the aberration of the accretion column to measure the expansion velocity of the wind at the position of the hot star. Again, the measured velocities confirmed a rather slow acceleration of the winds (Ahmad and Stencel 1988). From analyses of the C IV and Si IV emission lines and their phase dependence, Che-Bohnenstengel and Reimers (1986) also found evidence for a bow-shock region ahead of the secondary star in ζ Aur, 31 Cyg, 32 Cyg and 22 Vul. (A schematic representation of a bow shock is shown in Fig. 3.2.) At the same time, broad symmetrical emission lines from near-UV multiplets of Fe II in the much wider, but non-eclipsing, system δ Sge established evidence for an accretion disk (Reimers and Schröder 1983). The inference was that a typical ζ Aur system is too small for a stable accretion disk to form.

2.8.2 *Are the Wind and the Chromosphere a Single Structure or Two?*

The Sun's outer atmosphere is very highly structured, and is dominated by magnetic loops and the ejection of material through open magnetic fields that originate in the photosphere. The relationship between its chromosphere and wind is exceedingly complicated; suggestions for feeding the wind range from the diffusion of neutral material from spicules into magnetically open regions (Fontenla et al. 2002; Judge 2008) to open regions denuding active regions through magnetic recombination (Yang et al. 2013; Galinski and Shevchenko 2013).

For giant and supergiant stars it is generally assumed that the chromosphere and the wind are a single structure—in effect, a wind that proceeds directly from the whole surface of the star. But while observations of the ζ Aur binaries, such as the *HST* spectra analyzed by Baade et al. (1996), suggest that the wind is quite anisotropic, observations of Hα show that the chromosphere must cover the surface of such a star rather uniformly. Just how different are the processes that feed the ζ Aur-type winds on the one hand, and the solar wind on the other?

Connected to this question of the relationship between chromospheres and winds is the uncertainty of what the high electron densities that are deduced from balancing photoionization and recombination actually mean. If they imply microscopic clumping, which is the traditional thinking, then the chromosphere and the wind could well be a single structure that is distributed evenly over the stellar surface, as in the empirical model of Eaton (2008), though as yet unidentified forces must then be invoked to produce the clumping. On the other hand, if the chromosphere and wind consist of magnetically confined formations as in the solar atmosphere, they could well be separate structures that are anchored to mutually exclusive patches of the photosphere, and the need for microscopic clumping is then obviated. The uncertainty over which is the right model is not confined to ζ Aur primaries, though in the case of single stars it is likely to be far less constrained since increased uncertainties in the parameters of the models offer even more extreme—though possibly more unacknowledged—opportunities to disguise uncertainty in the structure.

Despite the similarity in the chromospheric properties of 22 Vul and ζ Aur, the two supergiants differ substantially as regards the properties of their winds. Schröder and Reimers (1989) found that the ionization of the wind of 22 Vul is very inhomogeneous. The primary has a slightly higher surface gravity than ζ Aur, and—as Schröder (1990) emphasized—it lies right on the Ca II cool-wind 'dividing line', suggesting that it is a border-line case. *IUE* spectra of Fe II lines in 22 Vul reveal incomplete wind line profiles, indicative of an absence of Fe II from whole sectors of its wind, whereas in ζ Aur (according to Che et al. 1983) Fe II remains a good proxy for the total density throughout the whole of the wind region.

2.9 The ζ Aur Primaries as Templates for Single Stars

Binary stars yield astrophysical information such as stellar mass, radius and evolutionary state with greatly superior precision than is routinely possible in the case of single stars, and in that regard the ζ Aur binaries emerge supreme because the orbital inclination is also known and therefore the masses given by the orbit solution are actual, rather than minimum, values. Moreover, since the radii and luminosities of the component stars can also be derived particularly well from a combination of photometry and spectroscopy, matching their positions in the H–R diagram to evolutionary tracks calculated for their corresponding masses provides a very valuable check on the parameters and assumptions which the evolution theory has adopted.

Most of the ζ Aur binaries can justifiably be described as 'detached' to a first-order approximation, and the mostly good levels of agreement between their positions in the H–R diagram and the predictions from stellar evolution prepared by Schröder et al. (1997) would support that general description; certainly they uncovered no strong indication that stars in binaries evolve differently from single ones. The model chromospheres that are produced by analyzing atmospheric eclipses of ζ Aur stars should therefore be representative of other similar cool but single (super)giants, and high interest in the ζ Aur stars has been maintained for nearly a century on the strength of that very premise. But it was a legitimate concern over the validity of such widespread assumptions that prompted the questions in Sect. 1.8, and it is a concern that cannot be dismissed lightly.

The tenet that binaries have unique potential as tracers of stellar evolution rests on the proviso that the evolution of a member of a binary system is not affected by its being gravitationally bound to another star, however distant. Just how realistic is that proviso? To what levels of refinement do the results derived for these ζ Aur giants, particularly the semi-empirical models of chromospheric structure and of the acceleration of the wind, represent the conditions that actually obtain in the outer atmospheres of single stars? Are the mass-loss rates and wind structures equivalent? Are the chromospheres in any way compatible? Or—expressed a little differently—are the lines from the outer winds from a *single* star thick enough to define the structure of that wind, and just where do the wind diagnostics of ionization cease to apply in the case of (say) Ca II because its levels of ionization change?

These fundamental questions have been raised frequently in recent years as more stellar objects prove to be exceptions to some norm and the validity of statistically-derived quantities is increasingly eroded away. The reviews and discussions in this book cannot, by themselves, provide definitive answers; the sample of objects is too small and the current empirical evidence much too patchy, so the products of the research to date only constitute a step along the road—and we cannot dictate where that road will lead. Moreover, studies of single stars require different approaches and methodologies, and those presently lag behind the sophistication of the tools of the binary-star analyst. What would therefore be more useful at this stage is to search for commonalities among the specific.

In many respects the chromosphere of ζ Aur does seem to be broadly representative of what is known about single K-supergiant chromospheres: the geometrical extent is comparable to the giant's radius, a wind is slowly accelerated through the outer regions, hydrogen remains mainly neutral, and Fe I and Ti I are the main sources of the absorption-line spectrum in the very lowest layers—though Fe may be predominantly singly ionized even there, so if neutral species are adopted as density tracers then the ionization balance must also be included.

There is proof, however, that the hot star in ζ Aur influences at least the outer layers of its primary's chromosphere. The phase-dependent emission that has been observed in Si I λ 3905 Å (Sect. 2.4.1.1) is clear evidence that the B-star's flux causes local over-ionization that creates a sub-stellar hot-spot in the cool star's outer chromosphere during the half-cycle of phases centred on periastron. Other analyses demonstrate that the ionization of Fe I to Fe II is nearly complete (Fe I / Fe II $<10^{-3}$), and that K-line absorption is noticeably reduced by photo-ionization, and moreover that those effects are more pronounced in the outer chromosphere than lower down, suggesting an external rather than an internal cause. Nevertheless, that B-star flux is not energetic enough to ionize hydrogen, even in the outer chromosphere; it is found that the B star has only a rather small Strömgren sphere, so the temperature, electron densities and total density of the chromosphere seem to be largely unaffected *as a whole* by the presence or nearness of the hot companion. It can therefore be inferred that the degree of ionization of hydrogen reflects the local heating mechanisms which are active in the chromosphere.

Nevertheless, the fact that the winds of these binary systems accelerate slowly seems at odds with the situation in cool *single* stars, where analyses of emission-line profiles usually yield accelerations that are characteristically much faster. One difficulty is that the wind diagnostics used for measuring mass loss and acceleration in binaries on the one hand and in single stars on the other are mutually exclusive. For single stars, intrinsic chromospheric emissions are scattered by overlying wind lines, whereas in binaries those lines are smothered by the light of the hot star that is scattered by the wind. Against that, in binaries that very spectrum of local scattering *from* the line of sight and global scattering *into* the line of sight yields a reliable measure of the wind kinematics, but comparable observations simply cannot be made for single stars.

This apparent discrepancy places doubt on the hypothesis that the cool components of ζ Aur binaries can be regarded as true surrogates for single stars. Eaton (2008) argued that they can and that it is the measurements of *single* stars that are discrepant, and offered a simplified model to show how that could be the case. Trying to judge the similarity of chromospheric structures in different objects is difficult in principle because in binaries most of the line emissions seen in total eclipse come from light scattered into the line of sight by the wind, so those chromospheres cannot be analyzed with the same semi-empirical models that can be used for single stars.

There are, however, a few lines that are intrinsically weak enough to be excited directly by the chromosphere, including inter-system lines of common elements and a few fluorescent lines. The surface fluxes of the inter-system lines Al II] and C II], and of certain fluorescent Fe I lines, seem to be about the same in binaries as in single stars (Schröder et al. 1988; Eaton 1992; Harper et al. 2005). Whether the C II] multiplet ratios give the same electron densities, though, is difficult to determine since the relevant *HST* spectra have not been observed. Those line-ratio measurements would have been particularly valuable for 31 Cyg, in which line-scattering is sufficiently weak that reliable measurements could probably have been achieved for C II]; in ζ Aur they were totally swamped by wind-scattered emission (Harper et al. 2005), and would have been even more so in 32 Cyg.

Another possibility is to use the Balmer lines. Hα profiles seem to be very similar both in single K supergiants and in the cool components of ζ Aur binaries (e.g., figs. A8–A21 of Eaton 1995), suggesting that the chromospheres probably do have approximately the same mass and velocity structure.

2.10 Looking Ahead

In order to foster a more detailed understanding of the physics which the ζ Aur binaries offer, more data are needed for interpretations that are based on theoretical models of the chromosphere, though the opportunities to advance understanding through that route will of course depend on the kinds of observing that is done. While ample archival material is still available from space missions, valuable new data for these bright stars could be gathered by systematic observing campaigns, possibly operating in robotic mode, using ground-based high-dispersion spectrographs capable of detecting the near UV.

Spectra of the optical and near-UV regions continue to offer distinct advantages. In the densest parts of the chromosphere—the inner few tens of a solar radius—the material is so massive that space-UV spectra are too saturated to be optimally useful. Changes in the apparent structure of the chromosphere and wind could be monitored from eclipse to eclipse, and from ingress to egress in the same eclipse, through near-UV spectra, as is already being attempted for some of the systems. Deriving more precise measures of that variability will provide clues to the nature of the inhomogeneity of the gas, its possible confinement by magnetic fields, and how its structure may change with height.

Very considerable progress will also be achieved by a set of synergistic observing programmes that involve other technologies to complement the conventional ground-based or space systems that supply more traditional types of spectra. The use of robotic telescopes could be promoted to support analyses of the light variability of these binaries. Monitoring a system throughout its orbit would offer information about proximity effects, ellipsoidal variations and reflection effects, but those kind of observations would be greatly enhanced if conjoined with asteroseismic observations to detect and define any pulsations in the cool supergiant; as emphasized in Sect. 7.3, eclipsing binaries are becoming a central feature in asteroseismology. Networks of robotic telescopes could refine the timings of ingress and egress at each eclipse, thus not only placing a system's geometry on a firmer footing but also providing evidence of variations in chromospheric mass, or changes in radius of the cool component brought about by pulsations. Interferometric imaging (Chap. 6) is another innovative route to explore, just as it helped to settle a decades-long debate over the nature of the secondary component in ϵ Aur (Chap. 4). Polarimetric observations, particularly if pursued throughout an orbital cycle, will be of considerable assistance in distinguishing material that is transferring mass from material that is regularly present in the system. New opportunities could be investigated to observe ζ Aur systems at radio frequencies, as carried out by Harper et al. (2005), in order to examine assumptions about how the observed gas is distributed around the cool supergiant. The final sum will certainly be richer than the sum of the individual parts if that synergy is well planned and orchestrated.

One of the biggest challenges—and opportunities—of the coming years is to calculate detailed semi-empirical models starting from certain assumptions about the distribution of gas and velocities as observed in specific systems, and by building on the pioneering attempts of Marshall (1996), Baade et al. (1996), Eaton (2008) and many others. One starting point of particular interest would be to use the Balmer lines to measure the excitation of hydrogen at different levels in the chromosphere, in order to place stronger constraints on the distributions of gas and Lyα radiation which the models adopt.

Opportunities must be pursued to bring theory and observation face to face more often and more thoroughly. As was mentioned in Sect. 1.5, since the physical parameters of the ζ Aur systems are mostly known quite precisely, particularly the masses of the cool evolved components, those binaries constitute a valuable set for testing stellar-evolution theory. Limitations are mostly brought about through uncertainties in the mass of the secondary star, and sometimes through which metallicity to adopt. Yet though Schröder et al. (1997) were able to carry out such tests on ζ Aur systems with considerable exactitude, one arrives at the frustrating conclusions that no two giants are fully alike, and that therefore no one current theory can adequately fit them all when the data are of sufficiently high quality. Is that an actual limitation, or is it a shortcoming that can be outpaced? Therein lies a number of challenges for the future.

Appendix: Digitizing Photographic Spectra

Transforming a photographic observation into an electronic file requires some design of adequate digitizing microdensitometer (nowadays a rare possession of any observatory), plus the trained expertise to manage photographic data correctly (also a rare commodity in an all-digital community).

Digital technology and tools for scanning photographic plates were being introduced into astronomy in the 1960s, though were still limited to paper-chart output and arduous manual tasks. A major breakthrough came with Photometric Data Systems' development of a digitizing microdensitometer, known as the PDS. The PDS worked at high speed and high spatial resolution. Operating in a step-wise sequence controlled by absolute encoders, it measured digitally the light transmitted through well-screened apertures below and above the emulsion and onto a photomultiplier, and wrote the output to magnetic tape, the then standard for digital data. PDS machines were acquired by a number of astronomical observatories, mostly (though not exclusively) in North America and Europe. They were latterly marketed by the American multinational technology company PerkinElmer, which bought out PDS in 1973. The worldwide adoption of digital observing and data management practices triggered the eventual disappearance of the PDS (and the attendant expertise) from most observatories. Fortunately the DAO has maintained, and has from time to time upgraded, the PDS which it acquired in 1979.

There is often confusion over approaches to 'scanning' astronomical plates. In common parlance a 'scanner' is a commercially-built desk-top instrument—general purpose, available off-the-shelf, and requiring little training to operate. But does it produce the same results as a PDS? If the answer were an emphatic 'Yes', then astronomy would have already digitized the entire contents of all of its plate archives. But the desk-top scanner was not designed to cope with the quantitative requirements of astronomical observations; its objective was to make a visually pleasing reproduction of a sample, be that sample a photograph, a letter or book, or (nowadays) a 3-D object. Imaging a photographic plate involves illuminating it in some way, and if that is done by shining a light either on or through the whole (or even strips of it), the emulsion will scatter the light and thereby vitiate the representation of the photometric information recorded by each pixel of the detector. The degree of scattering will also depend upon the density of blackened grains in the neighbourhood of each recorded pixel, and so cannot easily be modelled. A commercial scanner is also much less rigidly built than a microdensitometer of PDS stature, and is liable to positional errors, particularly near the edges of a sample. Although positional errors are of more minor consequence for spectra (which are the prime concern here), they can have a deleterious effect for direct plates, particularly if required for astrometry. However, the desk-top scanner is a very handy instrument for generating 'quick-look' scans of a plate in order to display what it contains, or to image log-book pages so as to share what they show without handling fragile originals.

The PDS, on the other hand, *was* designed to digitize astronomical plates correctly. It is sturdily built (models optionally rest on a metal or granite pillar), and a pre-slit screen eliminates light from all directions except from directly below the analyzing slit. The latter narrows the beam further to the desired width. The light that passes through the emulsion and onto the photomultiplier at each encoded step position is therefore an unambiguous measure of the density of the emulsion at just that step, and should be uncontaminated by scattering. As with any high-precision optical instrument, major enemies of performance are temperature fluctuations and dust, so a clean, controlled environment is important. The stability of the illuminating lamp is also critical, and needs a stabilized power supply that is protected from local interference.

Upgrades to the DAO's PDS over the years have incorporated various then-current state-of-the-art technologies, and include increasing its output resolution from the original 8 bits to 12, enhancing the positional encoders with a laser interferometer system, driving it from a dedicated PC, and sending the output to disk as FITS files.

The price of thoroughness is slow productivity. The PDS can work at a speed that is rapid enough to scan the exposed regions of (say) three or four 8-in. spectrograms per hour, but that is also approximately the time taken to process the output through a semi-automatic pipeline that extracts 1-D spectra fully linearized in both wavelength and intensity, so the operation is carried out in its entirety without necessarily creating bottlenecks. Some 100 spectra each of ζ Aur, VV Cep, 31 Cyg and 32 Cyg from the DAO archive have been digitized in this way, and made available (as FITS files) on the CADC public server at www1.cadc-ccda.hia-iha.nrc-cnrc.gc.ca/en.

References

Ahmad, I.A.: ApJ **301**, 275 (1986)
Ahmad, I.A.: ApJ **338**, 1011 (1989)
Ahmad, I.A., Chapman, R.D., Kondo, Y.: A&A **126**, L5 (1983)
Ahmad, I.A., Parsons, S.B.: ApJ, **299**, L33 (1985)
Ahmad, I.A., Stencel, R.E.: ApJ **329**, 797 (1988)
Ake, T.B., Parsons, S.B, Kondo, Y.: ApJ, 298, 772 (1985)
Ake, T.B., Barksdale, W.S., Fried, R.E., et al.: IBVS **2847** (1986)
Ake, T.B., Parsons, S.B.: IBVS **3002** (1987)
Ake, T.B., Parsons, S.B.: in A Decade of UV Astronomy with the *IUE* Satellite **1** (ESA SP 281a, 1988), p. 213
Ake, T.B., Robinson, R.: in Cool Stars, Stellar Systems, and the Sun, ed. by F. Favata, G.A.J. Hussain, B. Battrick (ESA SP-560, 2005), p. 381
Baade, R.: A&A **233**, 486 (1990)
Baade R., Kirsch, T., Reimers, D., et al.: in Cool Stars, Stellar Systems, and the Sun, ed. by R.J. Garcıa López, R. Rebolo, M.R. Zapaterio Osorio, M. Rosa (ASPC **223**, San Francisco, 2001), p. 1585
Baade R., Kirsch, T., Reimers, D., et al.: ApJ **466**, 979 (1996)
Bauer, W.H., Stencel, R.E.: ApJS **69**, 667 (1989)
Bauer, W., Bennett, P.B.: ApJS **211**, 27 (2014)
Bennett, P.D., Brown, A., Linsky, J.L.: ApJ **455**, 317 (1995)
Bennett, P.D., Harper, G.M., Brown, A., et al.: ApJ **471**, 454 (1996)
Brown, A., Bennett, P.D., Baade, R., et al.: AJ **122**, 392 (2001)
Buss, R.H., Snow, T.P.: ApJ **335**, 331 (1988)
Chapman, R.D.: ApJ **248**, 1043 (1981)
Chapman, R.D.: BAAS **15**, 925 (1983)
Che, A., Hempe, K, Reimers, D.: A&A **126**, 225 (1983)
Che-Bohnenstengel, A., Reimers, D.: A&A **156**, 172 (1986)
Christie, W.H., Wilson, O.C.: ApJ **81**, 426 (1935)
Dhital, S., West, A.A., Stassun, K.G., et al. AN **334**, 14 (2013)
Di Benedetto, G.P., Ferluga, S.: A&A **236**, 449 (1990)
Doherty, L.R., Jung, A.F.: ApJ **195**, 121 (1975)
Drake, S.A., Linsky, J.L.: AJ **91**, 602 (1986)
Drake, S.A, Brown, A., Reimers, D.: in Cool Stars, Stellar Systems, and the Sun, ed. by J.L. Linsky, R.E. Stencel (Springer, Berlin, 1987) p. 322
Dupree, A.K., Reimers, D.: in Exploring the Universe with the *IUE* Satellite, ed. by Y.Kondo, ASSL **129** (Reidel, Dordrecht, 1987), p. 445
Eaton, J.A.: ApJ **333**, 288 (1988)
Eaton, J.A.: MNRAS **258**, 473 (1992)
Eaton, J.A.: ApJ **404**, 305 (1993a)
Eaton, J.A.: AJ **105**, 1525 (1993b)
Eaton, J.A.: AJ **106**, 2081 (1993c)
Eaton, J.A.: AJ **109**, 1797 (1995)
Eaton, J.A.: in Cool Stars, Stellar Systems, and the Sun, ed. by R. Pallavicini, A.K. Dupree (ASPC **109**, San Francisco, 1996), p. 503
Eaton, J.A.: AJ **136**, 1964 (2008)
Eaton, J.A., Bell, C.: AJ **108**, 2276 (1994)
Eaton, J.A., Henry, G.W., Seeds, M.A.: in Cool Stars, Stellar Systems, and the Sun, ed. by J.-P. Caillault (ASPC **64**, San Francisco, 1994), p. 696
Eaton, J.A., Shaw, F.G.: AJ **133**, 2669 (2007)
Eaton, J.A., Henry, G.W., Odell, A.P.: ApJ **679**, 1498 (2008)
Fontenla, J.M., Avrett, E.H., Loeser, R.: ApJ **572**, 636 (2002)

Fox, G.K., Griscom, L.: MNRAS **278**, 975 (1996)
Galinski, V.L., Shevchenko, V.I.: ApJ **763**, 31 (2013)
Gray, D.F.: AJ **135**, 1450 (2008)
Griffin, R.& R.:JA&A **7**, 195 (1986)
Griffin, R.& R., Lines, H.C., Lines, R.D., et al.: IAPPP Comm. **44**, 82 (1991)
Griffin, R.E.M.: JRASC **82**, 49 (1988)
Griffin, R.E.M.: in Binary Stars as Critical Tools & Tests in Contemporary Astrophysics, ed. by W.I. Hartkopf, E.F. Guinan, P. Harmanec (IAUS **240**, Cambridge, 2007), pp. 645, 700
Griffin, R.E.M., Griffin, R.F., Schröder, K.P., et al.: A&A **234**, 284 (1990)
Griffin, R.E.M., Schröder, K.-P., Misch, A., et al.: A&A **254**, 289 (1992)
Griffin, R.E.M., Hünsch, M., Marshall, K.P., et al.: A&A **274**, 225 (1993)
Griffin, R.E.M., Marshall, K.P., Griffin, R.F., Schröder, K.-P.: A&A **301**, 217 (1995)
Griffin, R.E.M., Stencel, R.E.: PASP **125**, 775 (2013)
Griffin, R.F.: Sky & Tel **81**, 598 (1991)
Griffin, R.F.: Observatory **125**, 1 (2005)
Griffin, R.F.: Observatory **128**, 362 (2008)
Griffin, R.F., Griffin, R.E.M., Snyder, L.F., et al.: IAPPP **57**, 31 (1994)
Groth, H.G.: Zeitschr.f.Ap. **43**, 185 (1957)
Guinan, E.F., McCook, G.P.: PASP **91**, 343 (1979)
Guthnick, P.: AN **252**, 13 (1934)
Hack, M: A&A, **99**, 185 (1981)
Hack, M., Stickland, D.: in Exploring the Universe with the *IUE* Satellite, ed. by Y.Kondo, ASSL **129** (Reidel, Dordrecht, 1987), p. 445
Harper, G.M., Wilkinson, E., Brown, A., et al.: ApJ **551**, 486 (2001)
Harper, G.M.: ApJ **720**, 1767 (2010)
Harper, G.M., Brown, A., Bennett, P.D., et al.: AJ **129**, 1018 (2005)
Hempe, K.: A&A **115**, 133 (1982)
Judge, P.: ApJ **683**, 87 (2008)
Kenyon, S.J.: The Symbiotic Stars, (Cambridge Univ. Press, Cambridge, 1986)
Kirsch, T., Baade R.: A&A **291**, 535 (1994)
Kirsch, T., Baade, R., Reimers, D.: A&A **379**, 925 (2001)
Kiss, L.L., Szabo, G.M., Bedding, T.R.: MNRAS **372**, 1721 (2006)
Koch, R.H., Pfeiffer, R.J.: AJ **87**, 1409 (1982)
Lee, E.K., Wright, K.O.: Publ. DAO **11**, 339 (1960)
Linsky, J.L., Haisch, B.M.: ApJ **229**, L27 (1979)
Marshall, K.P.: MNRAS **280**, 977 (1996)
McAlister, H.A.: AJ **86**, 1397 (1981)
Mullan, D.J., Stencel, R.E.: in Advances in Ultraviolet Astronomy, ed. by Y. Kondo, J.M. Mead, R. DeWitt Chapman (NASA CP-2238, Washington, 1982), p. 235
Osterbrock, D., Flather, E.: ApJ **129**, 26 (1959)
Parsons, S.B., Ake, T.B.: IBVS **2334**, 1 (1983)
Parsons, S.B., Ake, T.B.:, ApJS, 119, 83 (1998)
Parsons, S.B., Ake, T.B., Hopkins, J.L.: PASP **97**, 725 (1985)
Popper, D.M.: ApJ **134**, 828 (1961)
Reimers, D., Che-Bohnenstengel, A.: A&A **166**, 252 (1986)
Reimers, D., Schröder, K.-P.: A&A **124**, 241 (1983)
Reimers, D., Baade, R., Schröder, K.-P.: A&A **227**, 133 (1990a)
Reimers, D., Baade, R., Schröder, et al.: A&A **236**, L25 (1990b)
Savage, B.D., Jenkins, E.B.: ApJ **172**, 491 (1972)
Schröder, K.-P.: A&A **124**, L16 (1983)
Schröder, K.-P.: A&A **147**, 103 (1985)
Schröder, K.-P.: A&A **170**, 70 (1986)
Schröder, K.-P.: A&A **236**, 165 (1990)
Schröder, K.-P., Reimers, D., Carpenter, K.G., et al.: A&A **202**, 136 (1988)

Schröder, K.-P., Reimers, D.: A&A **208**, 223 (1989)
Schröder, K.-P., Griffin, R.E.M., Griffin, R.F.: A&A **234**, 299 (1990)
Schröder, K.-P., Griffin, R.F.: Sterne & Weltr. **30**, 296 (1991)
Schröder, K.-P., Hünsch, M.: A&A **257**, 219 (1992)
Schröder, K.-P., Griffin, R.E.M., Hünsch, M.: A&A **288**, 273 (1994)
Schröder, K.-P., Hünsch, M.: A&A **289**, 893 (1994)
Schröder, K.-P., Marshall, K.P., R.E.M. Griffin: A&A **311**, 631 (1996)
Schröder, K.-P., Pols, O.R., Eggleton, P.P.: MNRAS **285**, 696 (1997)
Snow, T.P., Buss, R.H., Gilra, D.P., et al.: ApJ **321**, 921 (1987)
Stencel, R.E., Kondo, Y., Bernat, A.P, et al.: ApJ **233**, 621 (1979)
Stencel, R.E., Hopkins, J.L., Hagen, W., et al.: ApJ **281**, 751 (1984)
Taylor, A.R., Seaquist, E.R.: ApJ **286**, 263 (1984)
Toner, C.G., Gray, D.F.: ApJ **334**, 1008 (1988)
Unsöld, A.: Physik der Sternatmosphären (Springer, Berlin, 1968), p. 290
Wellmann, P.: Veröff. Univ. Berlin-Babelsberg **12**, No. 4 (1939)
Wilson, O.C.: ApJ **107**, 126 (1948)
Wilson, O.C.: JRASC **51**, 70 (1957)
Wilson, O.C., Abt, H.A.: ApJS **1**, 1 (1954)
Wright, K.O.: Publ. DAO **9**, 189 (1952)
Wright, K.O.: Publ. DAO **11**, 77 (1959)
Wright, K.O.: Vistas Astr. **12**, 147 (1970)
Wright, K.O., Huffman, R.E.: Publ. DAO **13**, 275 (1968)
Wright, K.O., Lee, E.K.: Publ. DAO **11**, 59 (1959)
Wright, K.O., Odgers, G.J.: **JRASC**, 56, 149 (1962)
Yang, L., He, J., Peter, H., et al.: ApJ **770**, 6 (2013)

Chapter 3
The Special Case of VV Cephei

Philip D. Bennett and Wendy Hagen Bauer

Abstract VV Cephei (M2 Iab + B0–2 V) is the only known ζ Aur binary that has an M-type supergiant. Studies of its atmospheric eclipses provide a valuable extension to what can be learned by eclipse mapping, but are fraught with many challenges. The space-UV absorption spectrum is extremely rich, and line blending is severe. Wind interaction and accretion effects are prominent, and the spectrum of the B star companion is heavily veiled by overlying Balmer continuum emission, which usually exceeds that of the B-star's continuum flux in the UV. An accretion region immediately around the B star gives rise to a rich spectrum of lines, some of which sometimes show prominent inverse P Cygni profiles. The chromospheric absorption spectrum shows structure in the line profiles that persists throughout chromospheric eclipse, and implies the existence of large-scale supersonic, complex velocity flows across the whole extended chromosphere of the M supergiant. Spectrum formation in the atmosphere of VV Cep is more complex than in the other ζ Aur binaries owing to the much higher rate of mass loss from the primary. This chapter summarizes the fundamental stellar and orbit parameters of VV Cep, and presents key results found from three decades of UV astronomy. It also presents a simple model of the chromosphere and wind.

3.1 Introduction

The eclipsing binary VV Cephei (HR 8383, HD 208816) has an orbital period of 20.3 years. It has been classified variously, the consensus homing in on M2 Iab + B0–2 V, so the primary's spectral type is similar to that of α Ori. Estimates

P.D. Bennett (✉)
Department of Astronomy and Physics, Saint Mary's University, 923 Robie Street, Halifax, NS B3H 3C3, Canada

Eureka Scientific, Inc., 2452 Delmer Street Suite 100, Oakland, CA 94602, USA
e-mail: pbennett@ap.smu.ca

W.H. Bauer
Whitin Observatory, Wellesley College, 106 Central St., Wellesley, MA 02418, USA
e-mail: wbauer@wellesley.edu

T.B. Ake, E. Griffin (eds.), *Giants of Eclipse: The ζ Aurigae Stars and Other Binary Systems*, Astrophysics and Space Science Library 408,
DOI 10.1007/978-3-319-09198-3_3

of the spectral type of the hot companion range from O8, based on a mass of $\sim 20\,M_{\odot}$ given by a spectroscopic orbit (Hutchings and Wright 1971), to A0, based on the UV continuum observed with *IUE* (Faraggiana and Selvelli 1979; Hack et al. 1989). VV Cep is also the eponymous member of the 'VV Cep stars', which are M supergiants with early-type main-sequence companions. The VV Cep stars hold valuable potential for determining fundamental parameters of M supergiants (for which no accurate masses have been determined), though only VV Cep—the sole member that eclipses—can yield spatially-resolved information about the outer atmosphere and wind of such stars from observations made during its chromospheric eclipses.

VV Cep's orbit is represented in Fig. 3.1 with the orientation on the sky of $\Omega = 147°$ given by van de Kamp (1977). It is quite eccentric ($e = 0.346$), which strongly suggests that, despite many claims to the contrary, mass transfer has not occurred. Although accretion phenomena are associated with the B star in VV Cep, they are almost certainly caused by accretion from the dense wind of the M supergiant and not by Roche-lobe overflow. The binary is more complex than the other ζ Aur systems in various ways. The rate of mass loss from its M2 Iab primary is greater than for the other K- or G-type ζ Aur primaries, while the chromosphere and wind are so much more extended that the hot companion orbits deeply within it and its own spectrum is never seen, even in the UV—see Fig. 3.2. Instead, the hot companion appears to be immersed in a region of strong Balmer continuum emission coming from the vicinity of the hot star which partially, and variably, veils the spectrum of the stellar companion. We refer to the entity of the hot star plus the emitting regions immediately around it collectively as the 'hot companion'.

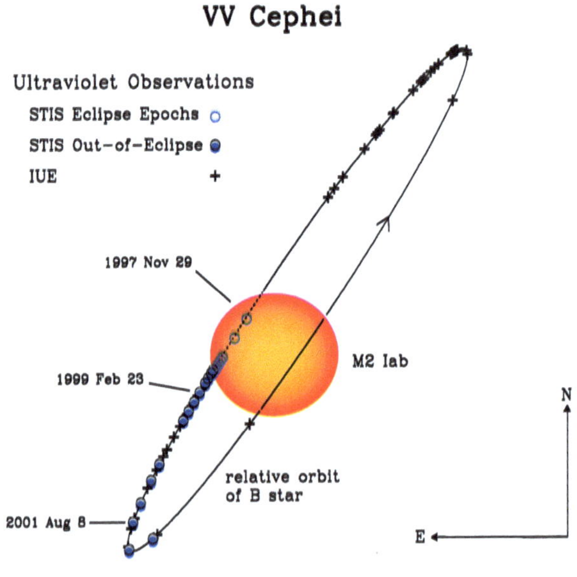

Fig. 3.1 The projected orbit of VV Cep, showing the phases of *IUE* and *STIS* observations

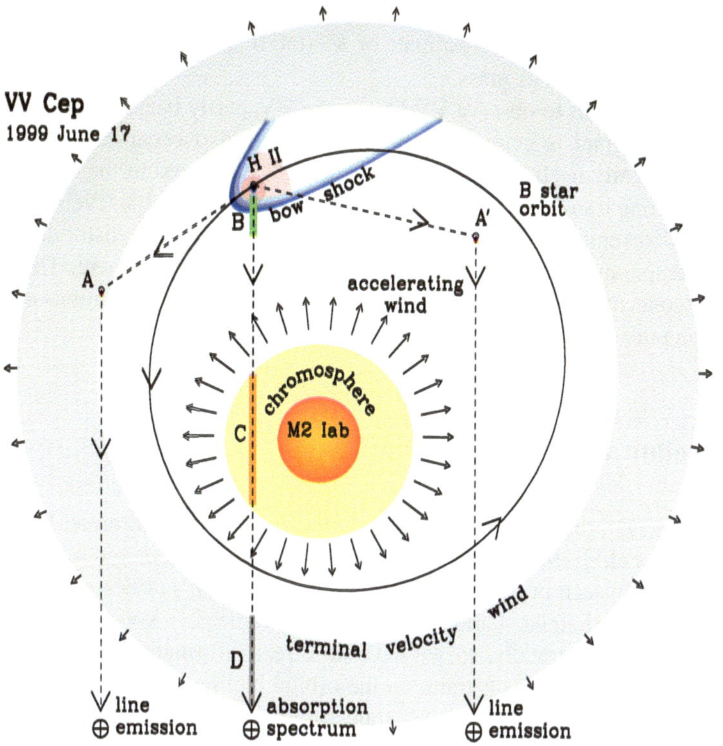

Fig. 3.2 Schematic diagram of VV Cep, drawn to scale, indicating where the various contributions to its observed UV spectrum are formed

A rich absorption spectrum similar to that of a Be shell star is seen whenever the hot companion is out of eclipse. It is therefore not possible to measure the RV of the B star by direct observation of its spectrum. Emission is seen in Hα; its RV varies in the opposite sense to the M-star's absorption lines, and moreover its intensity is greatly reduced during primary eclipse, implying that it originates near the B star. Wright (1977) measured the centroid of the Hα emission at different phases and derived an orbit for the secondary component on the assumption that his measurements represented the motion of the B star, though such an assumption may have been a little risky. The Balmer line and continuum spectrum seen in VV Cep is presumably a recombination spectrum from an H II region immediately round the B star, thereby corroborating its classification as early-B type. The peak of the Balmer-line emission depends both on the strength of the ionizing radiation from the B star and on the density of the circumstellar gas from the cool supergiant, so its centroid is likely to be biased towards higher circumstellar densities where n(H II) is greater. It probably does not, therefore, follow exactly the velocity displacements of the B star, but will be offset towards the system's barycentre. The RV of the Hα emission centroid may thus be biased in the direction of the centre of mass, so the observed shifts in the emission will be somewhat less than those of the B star, resulting in a systematic underestimate of K_2 (the B star's RV semi-amplitude) and thus an overestimate of its mass.

It is advantageous to observe VV Cep in the UV, partly because the contribution from the supergiant is negligibly small there, and also because there are many more scientifically useful spectral lines in the UV compared to the optical region, including strong transitions from ground or low-lying metastable levels. Metastable levels tend to remain populated, according to the Boltzmann distribution, at the sorts of temperatures and densities that pertain in a chromosphere. This chapter therefore concentrates on space-UV spectra, and chiefly on observations made with *HST* during the 1997–1998 eclipse.

3.2 Stellar and Orbit Solution

Wright's (1977) orbit solution, together with information from broad-band photometry, Hack et al.'s (1989) estimate of the visual extinction (derived from the size of the λ 2200 Å dip seen in *IUE* spectra) plus Leedjärv et al.'s (1999) eclipse timings, yield the orbit and stellar parameters set out in Table 3.1. VV Cep has not been resolved interferometrically, so we have no direct information about the angular size of the orbit. Further constraint on the orbital inclination, i, is needed in order to refine the latitude of the hot-star's trajectory behind the disk of the M star. The timings given by Leedjärv et al. and the eclipse chord derived from them is of no immediate help because only the *angular* diameter, and not the actual radius, of the M star, is known. Nevertheless, by taking those constraints together with the observed duration of an eclipse, some refinement has been achieved.

The values thus derived for a/R_1 and for $a_{\rm min}/R_1 = (a/R_1)(1 - e)$ (the minimum separation at periastron) indicate that the stars are well separated. For binaries of approximately equal mass ($q \approx 1$) and in Roche-lobe contact, the effective radius $R_{\rm L}$ of the Roche lobe is $R_{\rm L}/a = 0.38$ (Pacyński 1971; Eggleton 1983). Inverting that relation implies a minimum possible orbital separation of $a/R_{\rm L} = 2.6$ for stars in Roche-lobe contact, but in VV Cep that ratio is 3.3 and therefore significantly exceeds the limit. Even given the inherent uncertainties, it is still unlikely that the periastron separation in VV Cep is as small as the critical Roche-lobe separation of $a/R_{\rm L} = 2.6$. Moreover, binaries that are transferring mass are nearly always in circular orbits, and that is not the case for VV Cep. The evidence to hand therefore implies that VV Cep is not a mass-transfer binary, though the stars do approach closely at periastron. Better fundamental parameters for the system are needed in order to resolve completely this question of mass transfer.

The 1997–1998 eclipse occurred roughly 2 months later than predicted from the previously accepted period of 7430.5 days (Leedjärv et al. 1999). The cause of that behaviour is not known, but it is possible that radial pulsations in the M supergiant could skew the observed times of second and third contacts.

Table 3.1 Stellar and orbit parameters

Parameter	Symbol	Value
Period	P	7430.5 ± 20 days
Eccentricity	e	0.346 ± 0.01
Longitude of periastron	ω	$59^\circ.2 \pm 2^\circ.8$
M-star RV amplitude	K_1	$19.4 \pm 0.3\,\mathrm{km\,s^{-1}}$
B-star RV amplitude	K_2	$19.1 \pm 0.7\,\mathrm{km\,s^{-1}}$
Visual extinction	A_V	1.24 ± 0.2 mag.
M-star semi-major axis	$a_1 \sin i$	12.43 ± 0.20 AU
B-star semi-major axis	$a_2 \sin i$	12.24 ± 0.45 AU
M-star effective temperature	T_{eff}	3825 ± 100 K
M-star angular diameter[a]	θ_D	6.38 ± 0.6 mas
Inclination	i	$84^\circ.0 \pm 3^\circ.4$
Relative semi-major axis	a	24.81 ± 0.49 AU
Relative semi-major axis	α	16.2 ± 3.7 mas
M-star radius	R_1	$1050 \pm 250\ R_\odot$
M-star mass	M_1	$18.2 \pm 1.1\ M_\odot$
B-star mass	M_2	$18.6 \pm 1.1\ M_\odot$
M-star luminosity	L_1	$5.3 \pm 0.2\ L_\odot$
Distance	d	1.5 ± 0.3 kpc
Semi-major axis/M-star radius	a/R_1	5.1 ± 1.2
Periastron/M-star radius	a_{min}/R_1	3.3 ± 0.8
Epoch of mid-1997–1998 eclipse	T_0	JD 2 450 858
Eclipse duration		233 days
Eclipse chord length	R_{ecl}	7.3 AU

[a]Limb-darkened

3.3 Optical and Ultraviolet Spectroscopy

In addition to Wright's (1977) study of Hα throughout an orbit, the behaviour of emission in Hα and Hβ was also monitored by Möllenhoff and Schaifers (1981) during ingress into, and egress from, the 1977–1978 eclipse. The emission was double-peaked; the short-wavelength peak disappeared earlier than the other as the system went into eclipse and reappeared first as it emerged from eclipse, an observation that was interpreted as caused by the presence of a disk around the B star. Also working with optical spectra, Hack et al. (1992) included a study of the Balmer-line emission, while Kawabata and Saitō (1997) investigated forbidden-line features, from which they derived a mass-loss rate of $\sim 10^{-6}\ M_\odot\ \mathrm{year}^{-1}$.

Most of the new results since then have come from space-UV spectra. *IUE* was launched just in time to observe VV Cep during egress from the 1977–1978 eclipse, and early observations were published by Faraggiana and Selvelli (1979) and by Hagen et al. (1980). *IUE* observed VV Cep for 18 years—through nearly a full orbit—but unfortunately never in total eclipse in high-resolution mode, nor during eclipse ingress at all. Hack et al. (1989), Bauer et al. (1991), Stencel et al. (1993) and Bauer and Bennett (2000) followed *IUE* observations of it as far round the orbit as they could. Chromospheric absorption features from elements such as Fe I were observed to weaken and disappear as the system moved away from eclipse, but in all the *IUE* spectra the hot companion remained shrouded in a wealth of absorption lines. The continuum and the line profiles showed significant variations on timescales as short as 2 weeks; they were possibly related to variable accretion onto the hot companion.

However, while *IUE* taught us a great deal, it opened up mysteries too. At least some of those have since been understood through *HST* observations of total eclipse. Although it had been hoped that the contribution of the cool star to the spectrum would be minimal in the UV, it turns out that the rich emission-line spectrum which is observed during totality significantly affects the out-of-eclipse spectrum too.

The STIS spectrograph, installed on *HST* in 1997, observed VV Cep at 21 different epochs, beginning with a number during total eclipse and continuing past first quadrature. The observations were primarily made at high-resolution (~110,000) with the cross-dispersed echelle. Spectra at medium-resolution (~30,000) were also obtained, and low-resolution grating spectra in the near-UV and optical regions were observed through a large photometric aperture in order to provide accurate flux calibrations of the echelle spectra and to determine the spectral energy distribution of the hot companion.

Between 2000 and 2007 VV Cep was observed six times at shorter wavelengths by *FUSE*. At the shortest wavelength limit (λ 920 Å) the spectra show only airglow, but a continuum begins to come up around λ 1040 Å, and gives way longward to absorption features similar to those seen in 31 Cyg during deep chromospheric eclipse.

3.4 The Formation of Spectrum Lines in VV Cep

Longward of ~ λ 5000 Å the spectrum of VV Cep resembles that of an M2 supergiant like α Ori, except for the prominent Hα emission which rises to ~20 times the continuum level of the M star and indicates the presence of an H II region. Forbidden emission lines from iron-group ions are also visible in the optical spectrum. At somewhat shorter optical wavelengths prominent emission is seen in other Balmer lines, while shortward of ~λ 4000 Å the spectrum becomes composite too, as a (variable) Balmer continuum in emission from the hot companion can be seen. In the space-UV the M-star's contribution becomes negligible shortward of ~λ 3000 Å and the spectrum is dominated by that of the hot companion, so the luminosity of the system is largely that of the hot companion. However, the nature of the spectra that are observed also depends on the orbital phase.

As with the other ζ Aur binaries, the spectrum of VV Cep manifests three distinct states—normal (outside eclipse), chromospheric, and total eclipse—that are accompanied by highly characteristic changes. It is the details seen in those different states that mark VV Cep out as 'special'. Even the details are often confused by the strong blending that persists throughout the orbit.

During total eclipse the direct line of sight to the hot star is blocked, but photons from its continuum are scattered back into our line of sight by the extended wind of the M star and give rise to the rich emission spectrum that is observed. During chromospheric eclipse the spectrum would be expected to exhibit absorptions by the M-star's chromosphere superimposed upon the composite one; however, the variable nature of the emission associated with the hot companion provides an added set of complexities. Outside eclipse the line of sight probes the intra-binary gas, which is of substantially lower density than the chromosphere. Although some absorption can be identified as coming from that intra-binary material, the spectrum at those non-eclipse phases is dominated by the effects of highly-variable accretion in the immediate vicinity of the hot star.

Figure 3.2 depicts schematically where the different contributions to the emission may be located in or near the system. The orbit of the B star is shown, and on it the hot companion, an early B-type star, is surrounded by an accretion and shock region (the small filled blue circle below the H II label inside the bow shock region at the top). The direct line of sight shown from the hot star to the observer represents the orbital configuration on 1999 June 17. Superimposed on the UV continuum are features from the shock/accretion region (near point **B**), absorption from the M supergiant's chromosphere (point **C**), absorption from the accelerating wind (arrows crossing the line of sight), and absorption from the wind moving at its terminal velocity (filled grey region) in the foreground of the M supergiant (**D**). The arrows are schematic wind acceleration vectors, their lengths providing a qualitative estimate of the acceleration of the wind at those positions. The absorption line profiles seen along sight-line BCD suggest that there is no outflow in the chromosphere, but the wind appears to accelerate rapidly outside this region. That acceleration gradually tapers off beyond the radius of the orbit, where the wind attains its terminal velocity of $\sim$20 km s^{-1}. The emission-line spectrum, prominent during eclipse (but present at all times), is formed by photons from the hot companion that are scattered by atoms or ions in the supergiant's wind back into the observers's line of sight. Two such possible indirect lines of sight are shown.

Table 3.2 contains a log of the 21 *HST* spectra that were observed in the space-UV, mostly during the 1997–1998 eclipse, together with geometric quantities deduced from Table 3.1. These observations are now discussed within the three different states mentioned.

Table 3.2 *HST* observations of VV Cephei

Epoch	Program/ visit	UT date of mid-observation	Phase	R_p [a] (R_1)
1	7269/01	1997 Nov 29	0.990	0.74
2	7269/02	1998 Mar 22	0.005	0.696
3	7269/05	1998 Jul 15	0.020	0.839
4	7269/04	1998 Aug 1	0.023	0.874
5	7269/06	1998 Aug 20	0.025	0.913
6	7269/07	1998 Sep 8	0.028	0.957
7	7269/08	1998 Sep 23	0.030	0.99
8	7269/53	1998 Oct 14	0.033	1.04
9	7269/09	1998 Nov 1	0.035	1.086
10	7269/10	1998 Dec 2	0.039	1.165
11	7269/11	1998 Dec 29	0.043	1.238
12	7269/12	1999 Feb 23	0.050	1.391
13	7269/13	1999 Apr 21	0.058	1.55
14	7269/14	1999 Jun 17	0.066	1.714
15	8257/01	1999 Aug 12	0.073	1.877
16	8257/02	2000 May 19	0.111	2.665
17	8779/01	2000 Sep 7	0.126	2.955
18	8779/02	2001 Feb 22	0.149	3.352
19	9231/01	2001 Aug 8	0.171	3.683
20	9231/02	2002 Sep 2	0.224	4.089
21	9231/12	2003 Sep 25	0.276	3.724

Orbit points	UT date	Phase	R_p (R_1)
Mid-eclipse	1998 Feb 14	0	0.689
Quadrature	2002 Oct 6	0.228	4.093
Periastron	2005 Feb 3	0.343	1.783
Secondary eclipse	2005 Nov 24	0.382	0.373

[a] R_p, the 'impact parameter' in units of M-star radii, R_1, is defined as the point of closest approach of the line of sight to the centre of the M star.

3.5 Total Eclipse

The total-eclipse spectrum, which Bauer et al. (2007) plotted in its entirety, is rich in emission features. They identified the great majority of the lines, relying on the database of Hirata and Horaguchi (1995). As in other ζ Aur binaries, emission arises from scattering of the hot-star's photons by atoms and ions in the cool primary's wind, but in VV Cep the rate of mass loss by the cool primary is ~100 times greater than in those other binaries so the density in the wind is correspondingly larger. This gives rise to an emission-line spectrum that is very much stronger than in the classical ζ Aur stars (Fig. 3.3); it is *always* present in VV Cep—it is just less obvious when the system is out of eclipse.

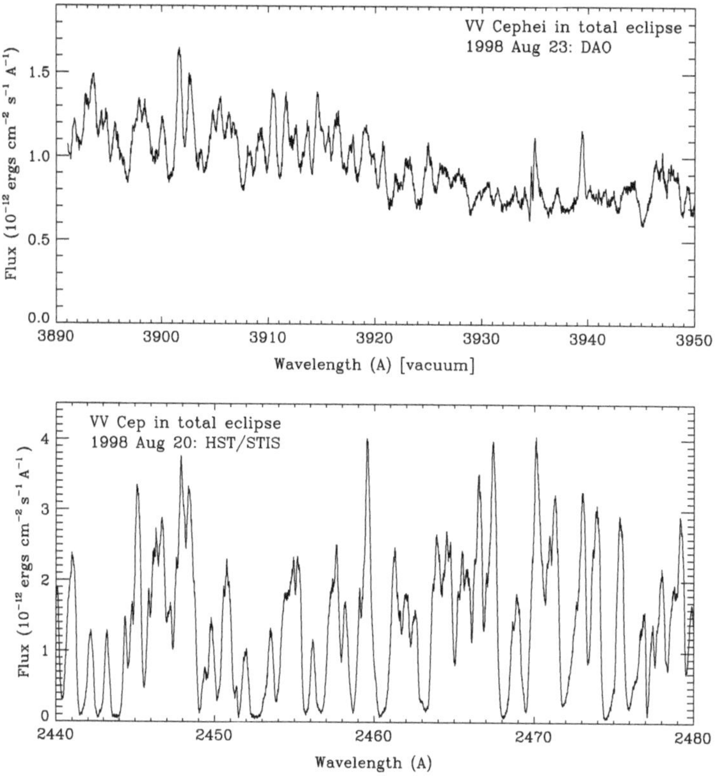

Fig. 3.3 Comparing optical and UV spectra during total eclipse. *Upper:* an optical spectrum, observed at the DAO on 1998 August 23, is a composite of the emission-line spectrum and the photospheric spectrum of the M star. *Lower:* a STIS spectrum (taken 3 days earlier) is very rich in emission lines. Both are shown in vacuum wavelengths

An intriguing new phenomenon seen for the first time in the eclipse spectrum was a weak continuum at short wavelengths. Near Lyα, that weak continuum strongly resembled what is seen far from eclipse but the flux was reduced by nearly an order of magnitude (Fig. 3.4). The eclipse continuum was therefore probably a scattered-light version of the out-of-eclipse spectrum. In view of the observed wavelength dependence, the source is probably Rayleigh scattering, which is dominated by the far wings of Lyα at the shortest wavelengths observable by *HST*.

The wind of the M supergiant has a terminal velocity of $\sim$20 km s^{-1}, as measured from blue-shifted absorption features (see Fig. 3.7). The wind is responsible for the emission-line spectrum that is seen during total eclipse, and influences the profiles of strong absorption features seen outside eclipse. The wind emission emanates from a spatially extended volume that is resolved in the STIS echelle spectra at all epochs observed. Bauer et al. (2008) employed a special reduction technique that enabled

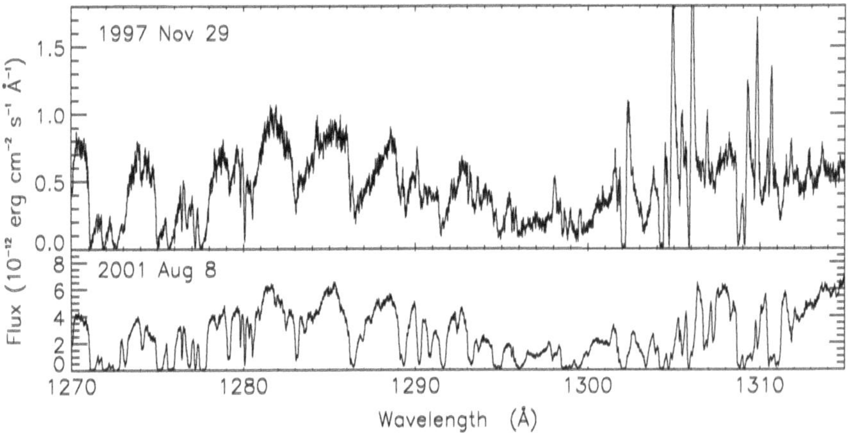

Fig. 3.4 Rayleigh scattering in the wind of VV Cep. Observation near mid-eclipse (*upper panel*), and near first quadrature (*lower panel*) (note the change in scale). Their close correspondence suggests that the total-eclipse spectrum results from scattering in the extensive wind of the M star. Sharp emission features observed during totality remain visible out of eclipse, and can influence an absorption profile, as seen here most easily in the line of P II at λ 1309.9 Å

them to identify which contributions to the emission lines originated in the extended region in the system, and which came from the unresolved stellar companion and its immediate surroundings (see Sect. 3.10).

3.6 Chromospheric Eclipse

In VV Cep the chromosphere appears to be a continuous extension of the photosphere. It is optically thin at optical wavelengths, but remains gravitationally bound to the star. It appears to have little in common with the solar chromosphere except for being a region that lies immediately exterior to the photosphere and is somewhat warmer than the underlying photosphere. The chromosphere of VV Cep is much more extended than that of the Sun; its full extent is about 1–1.5 R_1.

The UV continuum is opaque in the lower chromosphere owing to the very large opacity of the damping wings of Lyα and the bound-free photoionization cross-section of neutral metals (Fig. 3.5). As a consequence, lines of easily-ionized neutral atomic species such as Fe I, and of some singly-ionized atoms such as V II, that would otherwise be quickly ionized by the strong UV radiation field of the hot companion survive in the lower chromosphere where that UV radiation cannot penetrate. Because those neutral atoms or easily-ionized single ions are confined to regions close to the M supergiant, their chromospheric absorption-line profiles provide excellent diagnostics of the lower chromosphere.

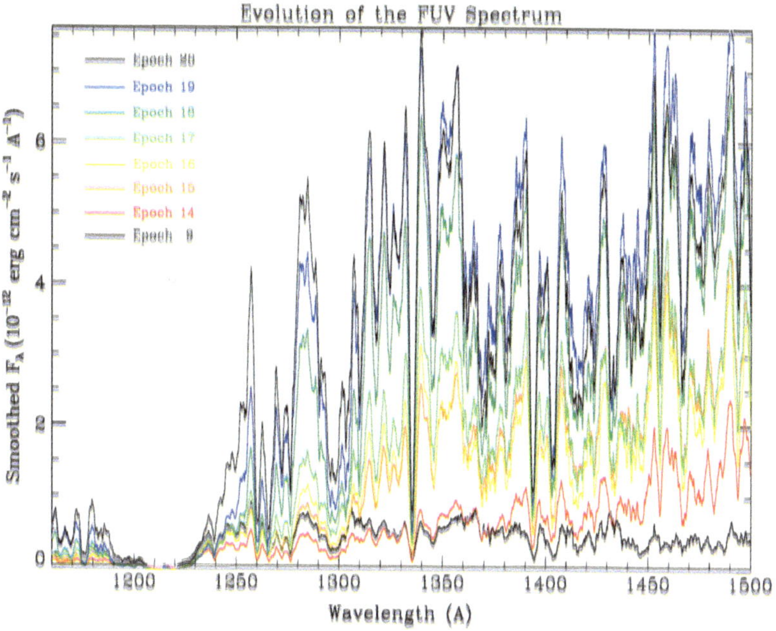

Fig. 3.5 The far UV spectrum emerges during egress from the 1997–1998 eclipse, as observed by STIS. The dates and orbital phases of the different epochs of the observations are listed in Table 3.2

Multi-component structure was seen throughout chromospheric eclipse, as is shown in Fig. 3.6. Fe I lines near λ 2995 and 3009 Å changed in strength as egress continued. By 2000 May 9 lines of most neutral elements, including Fe I and Ni I, had disappeared, although some lines of other neutral elements persisted longer. Chromospheric S I absorption near λ 1800 Å was also visible until 2000 Sep 7. A few elements remained neutral right through the outer atmosphere and into the wind. While some lines of singly-ionized elements weakened during chromospheric eclipse, they tended to persist longer than did lines of neutral species. The strong absorption feature near λ 3002 Å is a blend of several neutral lines and a 1.7-eV line of V II; by 2000 May 19 only the V II contribution could still be seen. Features from more highly-excited elements had also strengthened by then; the inverse P Cygni profiles of 4-eV Fe II and Cr II near λ 2998 and 3000 Å are noteworthy; they were very weak in the 1999 Aug 12 observation, but that was probably stochastic and related to varying accretion onto the hot star. The same lines were also seen in absorption on 1999 Feb 23.

The basics of the temporal evolution of the spectrum from mid-totality through to first quadrature are shown in Fig. 3.7; an emission-line spectrum observed during totality is plotted as dots. Weak emission features tended to be symmetric, while strong low-excitation lines were sculpted by blue-shifted absorption from the M supergiant wind. In the upper panel the total-eclipse spectrum is compared to the spectrum observed shortly after fourth contact; the strong absorption-line

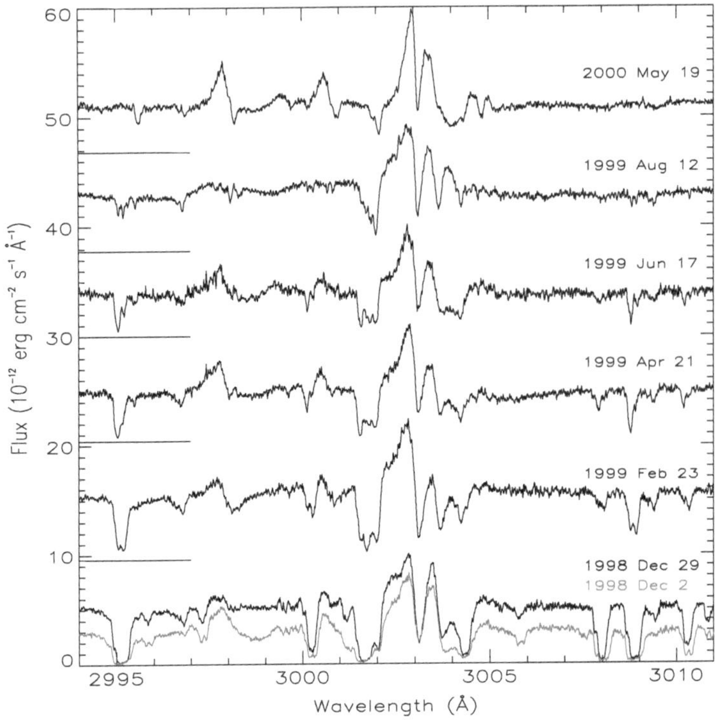

Fig. 3.6 Changes in line profiles during chromospheric eclipse. Fe I lines at λ 2995.3, 3000.4, 3008.2, 3009.0 and 3010.4 Å weakened and disappeared; they were double at all observed chromospheric-eclipse phases. By 2000 May 19, inverse P Cygni profiles of high-excitation features appeared, including an Fe II 4.49-eV line at λ 2998.2 Å and a blend of Cr II (4.0 eV) at λ 3000.8 Å and Fe II (3.8 eV) at λ 3000.9 Å. The strong double-peaked emission feature that appeared in all the observations is a 1.7-eV line of Fe II

profiles were modified by superimposed emission that occurred during total eclipse. The lower panel compares the total-eclipse spectrum to one observed near first quadrature. By that phase the strong V II lines had weakened and disappeared, as had two weaker lines of Ti I near λ 2943 and 2957 Å. The strong Fe II and Mn II absorption was still filled in by the emission that occurred during totality, although a shift in velocity was apparent. The wind absorption that was observed during total eclipse continued to be detected outside eclipse. There is a general correspondence between the emission seen during eclipse and the absorption seen out of eclipse. Some high-excitation absorption lines of Cr II and Fe II were visible, but were not phase-dependent. The same absorption was seen in some of the other early-egress observations, but was not visible in the one plotted in the upper panel.

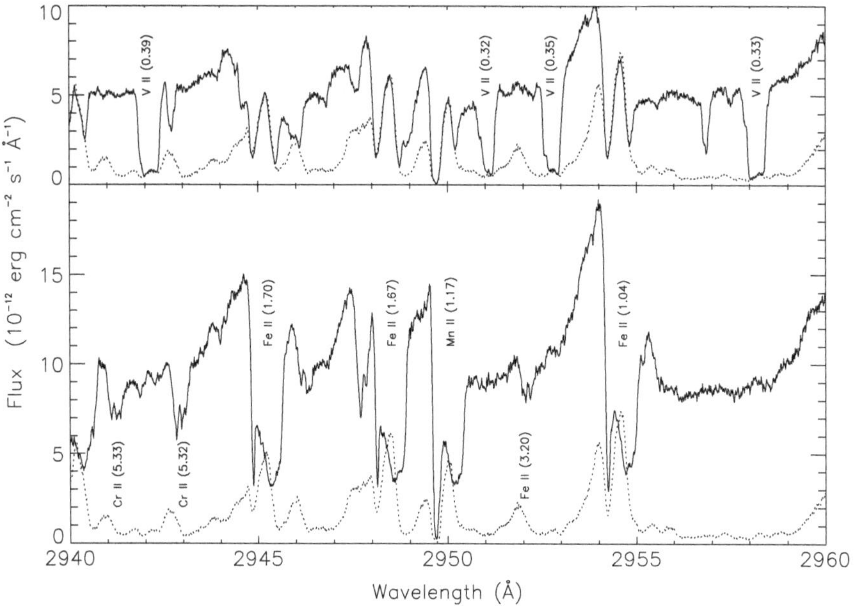

Fig. 3.7 Evolution of VV Cep spectra between total eclipse and first quadrature. The dotted spectrum in both is the totality observation of 1998 Nov 1. *Upper panel:* the first observation (1998 Dec 29) after the continuum had fully emerged. The strong Fe II and Mn II absorption lines were modified by the emission seen in the totality spectrum. *Lower panel:* an observation made near first quadrature (2001 Aug 8). Identifications and lower excitation potentials of some lines are marked

3.7 Chromospheric Structure Inferred from Line Profiles

Since UV spectra of VV Cep are uncontaminated by the M star, they are ideal for analyzing the details revealed in line profiles. Figure 3.8 illustrates the behaviour of two typical chromospheric absorption lines of different strengths. They are broad—a typical full-width at half-maximum (FWHM) is $\sim$25 km s^{-1}—and were resolved by STIS into two, and sometimes, three discrete components. The relative spacings of those components remained almost fixed at $\sim$17 km s^{-1} over the entire range of the chromospheric phases observed, with the M-star's RV lying roughly midway between them. But although the profiles were similar at all those phases, they shifted in velocity by about -10 km s^{-1} as egress progressed, as is seen clearly in the sequence of the Ti II λ 2913 Å line. The maximum velocity shift was reached on 1999 June 17, when the impact parameter was $R_\mathrm{p} = 1.71 R_1$ (Table 3.2). However, during those early egress phases there was almost no change in the M supergiant velocity, so the shift observed in the chromospheric lines was intrinsic to the chromosphere. As egress progressed and the sight-line to the hot component became more removed from the projected limb of the M star, the centroid of the chromospheric absorption line shifted back by about $+15$ km s^{-1}. Since the stellar

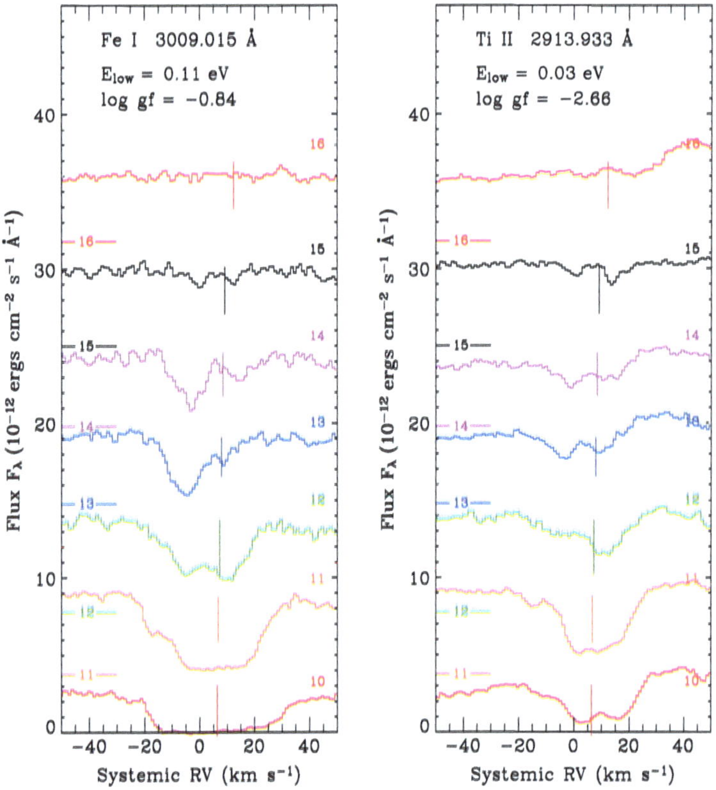

Fig. 3.8 The evolution of two prominent chromospheric absorption lines as the distance from the M star increases, plotted in the velocity rest-frame of the system. The column density responsible for absorption in the stronger Fe I line at λ 3009 Å (*left*) is consistent with that of the weaker Ti II line at λ 2913 Å (*right*). Vertical bars indicate the RV of the M star. Spectra are numbered by Epoch (Table 3.2); their zero levels are indicated by the horizontal lines on the left

RV had also shifted $+5\,\mathrm{km\,s^{-1}}$ between third contact and the end of chromospheric eclipse, the chromospheric line had in fact returned to the same velocity relative to the M star as it showed at third contact. The cause of that vagary in velocity is not understood, but is presumably related to the decoupling of angular momentum in the (rotating) M supergiant atmosphere, since the co-rotation that is apparent in the lower chromosphere must eventually decouple.

Column densities of absorbing gas along the lines of sight can be reconstructed from series of absorption-line measurements spanning an appropriate range of optical depths, by the 'apparent optical depth' method of Savage and Sembach (1991). The problem remains, however, of determining the location along the line of sight where a feature with a given RV forms. Constructing a 1-D spherically symmetric empirical model chromosphere requires the column density structure as a function of radial height, $N_{\mathrm{H}}(r)$, whereas the apparent optical depth method yields column densities as a function of wind velocity, $N_H(v)$. Those velocities do not vary monotonically with RV along the line of sight.

The shapes of the chromospheric line profiles and the separations of the components are consistent for lines of similar lower-excitation energy. For example, as Fig. 3.8 shows, the Ti II line at λ 2913 Å has a lower optical depth than does the strong Fe I λ 3009 Å line at most phases, but Ti II survives further out into the extended chromosphere: by 2000 May 19 the Ti II line is still weakly present but the neutral Fe I line has vanished. That disappearance thus marks the outer extent of the chromosphere. The entire velocity structure observed in these chromospheric profiles is therefore local to the chromosphere, because there is no significant column density contribution to the absorption features outside the radius corresponding to the impact parameter for 2000 May 19. The stellar wind also does not contribute to those chromospheric absorption components, so the velocity structure inferred from the doubled line profiles must be entirely confined within the chromosphere itself.

Results from analyzing these and other chromospheric lines strongly suggest that *all* of the complex velocity structure seen in the chromospheric absorption spectra is local in origin. The chromospheric profiles appear to reflect the velocity structure of the densest part of the chromosphere along the respective lines of sight, and do not result from a spatial velocity gradient like an accelerating wind. This is a perplexing result, because it means that discrete, supersonic velocity streams of different temperatures must coexist in the same persistent flow pattern along each line of sight through the chromosphere.

3.8 The Hot Companion

The out-of-eclipse spectrum of the circumstellar gas and wind is complex (Bauer and Bennett 2000), and a purely stellar spectrum of the hot star never fully emerges. Although some of the temporal chromospheric absorption features which also appear in other ζ Aur systems do weaken and disappear, other transitions which, in those other ζ Aur systems, eventually disappear at the conclusion of chromospheric eclipse remain in absorption in VV Cep throughout its orbit. They exhibit variable profiles, and variable emission is also superimposed throughout the orbit. Narrow absorption features, shifted by about $-20\,\mathrm{km\,s^{-1}}$, are formed in the M-star's wind. A small number of broad absorption features, such as resonance lines of Si IV and C IV, originate in the hot companion, and it was on that basis that Bauer et al. (1991) estimated a spectral type of B1–B2 for the hot star, although such lines could also originate in an accretion region rather than in a stellar photosphere.

The low-resolution STIS grating spectra yielded a new result: the UV continuum of the hot companion, which had long been suspected of having a non-stellar component, was found to make a substantial, and in some cases a dominant, contribution to the Balmer continuum. Figure 3.9 presents the intrinsic (de-reddened) spectrum of the hot companion at three phases: near third contact, mid-chromospheric eclipse,

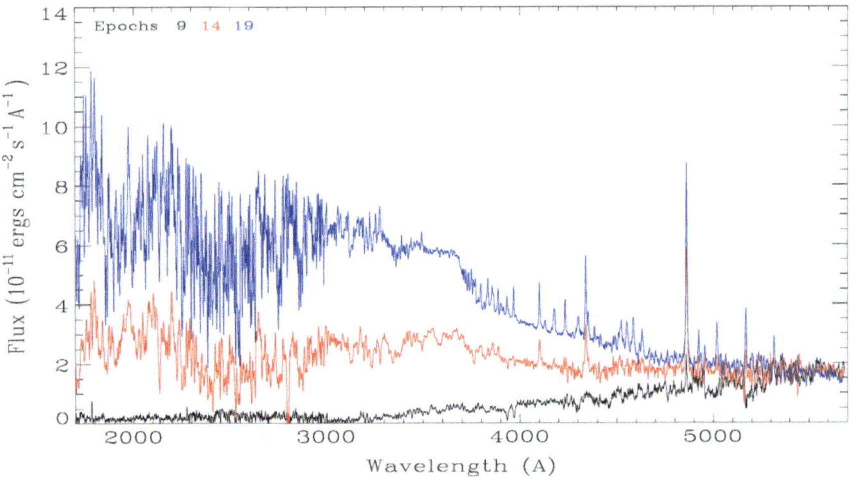

Fig. 3.9 De-reddened spectra of the hot companion at different phases, ranged upward by Epoch number. The Balmer emission was least on 1998 Nov 1 (Epoch 9) just after third contact, before the spectrum emerged in the UV. By 1999 Jun 17 (Epoch 14) the system was at mid-chromospheric eclipse, and by 2001 Aug 8 (Epoch 19, near quadrature) the UV continuum of the hot companion was at its brightest, Balmer continuum emission producing much of the observed UV continuum. Longward of λ 5500 Å the spectra converged to a common form, presumably that of the pure B star. The Balmer increment shortward of λ 4000 Å also became increasingly prominent as the sequence progressed

and out of eclipse. The spectra were created by subtracting from each observation a mid-eclipse spectrum, which was a composite of the spectrum of the M star itself and uneclipsed wind emission originating far from the hot star. The residue was then de-reddened to remove the effects of interstellar extinction, and should in principle have shown the pure spectrum of the hot companion. The Balmer emission from the hot companion was found to be highly variable, and on 2001 Aug 8 it was responsible for much of the observed UV continuum, when the higher Balmer lines to H14 were prominently in emission. Longward of λ 5500 Å the spectrum of the hot companion was stellar-like.

The inverse P Cygni profiles seen out of eclipse signal the existence of a wind accretion region immediately around the B star that provides a significant, and sometimes dominant, source of UV luminosity (Fig. 3.9). It also means that the 'bare' B-star spectrum is never seen. The very prominent Balmer emission associated with the shell spectrum suggests that the source is a recombination spectrum inside an H II region. That H II region must be of limited spatial extent because all the Balmer line and continuum emission vanishes during total eclipse except for Hα, though it weakens greatly.

The orbital motion of the B star through the M-star's wind is sufficiently supersonic that a weak bow shock must also be present around the hot star; it would enclose the H II region in a small 'bubble' inside it and immediately around the B star. The bow shock front (which moves with the B star) is expected to be responsible for the strong, slightly blue-shifted emission that is so prominent in the shell spectrum. Shortward of that emission peak there is a long, broad emission wing which—at times—extends to velocities up to $\sim$300 km s^{-1} from a line's centre. Such extreme velocities mean that those features must form in an in-falling accretion disk or stream close to the B star; no other spectral features in the VV Cep system show such high velocities. The substantial variability which the shell spectrum manifests may be caused by accretion from clumps within the M-star's wind.

Beyond the chromosphere, the stellar wind accelerates and the gas is no longer gravitationally bound to the M star. The wind is responsible for the formation of the UV emission spectrum that is seen in eclipse. Outside eclipse the evidence for the wind is predominantly as absorption components that are shifted by about -20 km s^{-1} in strong, low-excitation lines which presumably form in the foreground terminal wind velocity along the line of sight to the hot star, and therefore far from the M star. There is little evidence of a wind acceleration region, suggesting that the acceleration is either fast, with little column density, or intermittent and therefore not seen in all the observations. Any such region must lie outside the chromosphere, at a distance of at least R_1 from the surface of the M star.

3.9 Comparison from Orbit to Orbit

Egress of VV Cep has now been observed in the UV in two eclipses. In both 1978, seen with *IUE*, and 1998, seen with STIS, the continuum emerged later there than in the optical. While lines of neutral elements like Fe I and Ni I weakened and disappeared on a similar timescale on both occasions, there were significant differences in the levels of continuum flux, and in the weakening of ionized elements such as V II and Ti II that have low ionization potentials. Some strong V II and Ti II absorption features that were present immediately after egress in 1998 had vanished by first quadrature, although they had still been visible strongly at the same phase during the previous orbit. In the two *HST* observations in 2001 the level of continuum flux, ascribed to Balmer continuum emission, was significantly greater than was ever seen by *IUE*, even though the latter observed VV Cep during almost a complete orbit. That higher UV flux could have been responsible for destroying more of the easily-ionized ions such as V II, whereas in the previous orbit, when the flux level was lower, the same ions survived well out into the circumstellar environment.

On the other hand, the continuum was particularly low in the spectrum obtained on 2003 Sep 25. Although the calibration of that observation is somewhat suspect for technical reasons, it is worth noting that the faintest out-of-eclipse continuum measured during the *previous* orbit was at very nearly the same phase, so it is possible that the low continuum was in fact a phase-related feature.

The Mg II *h* and *k* lines behaved very differently between the two orbit cycles. Figure 3.10 illustrates those differences, commencing with partial eclipse in 1978 (coloured/thin) and in 1998 (drawn in black/thick) at closely corresponding phases. There were no significant changes in the two line profiles seen in the 1997–1998 eclipse between mid-totality (panel **a**) to almost the end of chromospheric eclipse (panels **b** and **c**), and also thus far resembled quite closely the profiles seen in the corresponding phases of the previous eclipse. Those profiles are characteristic of an expanding chromosphere and wind. However, by 1 year later (panel **d**) the shorter-wavelength emission peak of 1997–1998 had begun to strengthen compared to the previous orbit. A year after that, at first quadrature, in the earlier eclipse the profiles remained symmetrical with short- and long-wavelength peaks of approximately equal intensity, whereas in the later one the short-wavelength peak had become significantly stronger, reaching an extreme in the observation shown in panel **e**. A year after quadrature (panel **f**) the *h* and *k* profiles were once again quite similar in the two orbits. It is unlikely that the excess brightening of the peaks at quadrature in the later eclipse was related simply to the increases in the Balmer continuum emission that also occurred in that eclipse during the interval between egress and quadrature, because the continuum flux had in fact already declined by the time the *h* and *k* emission was strongest.

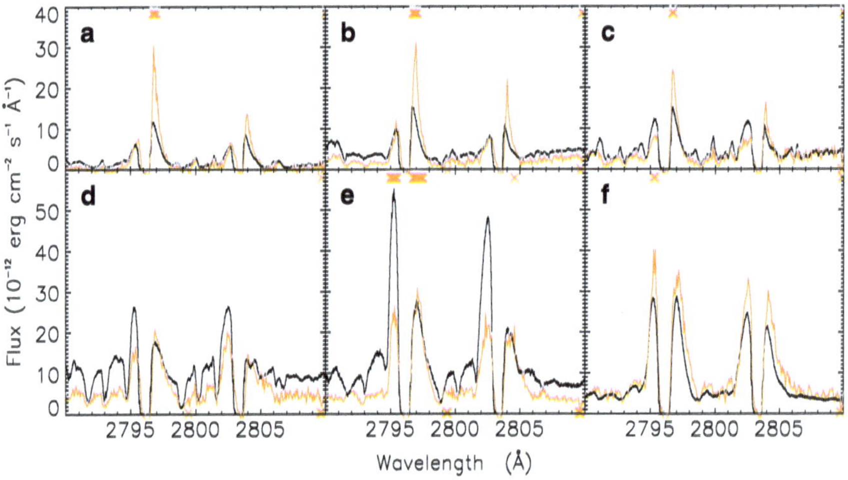

Fig. 3.10 Comparing the Mg II *h* and *k* lines as observed at approximately yearly intervals in the 1997–1998 orbit (drawn in *black/thick*) and at similar phases in the 1978 orbit (*coloured/thin*). In the latter, x represents a point with a non-zero quality flag; those at the bottom represent réseau marks which all *IUE* spectra bore, while most of those at the top are due to saturation caused by the bright features. The spectra in **a** correspond to a phase between third and fourth contact, **b–d** to egress, **e** to quadrature, and **f** to about a year later

3.10 VV Cep as an Extended Source

STIS was designed to make observations of extended sources.[1] The measured FWHM of the spatial point-spread function for standard stars is $\sim 0''.06$. Examination of the 2-D spectra of VV Cep indicated that portions of some features did not originate in a point source, but in a physically extended region (Bauer et al. 2008). The STIS spectra were therefore reprocessed in two dimensions with a spatial pixel size of $0''.0145$, corresponding to 4.5 times the M-supergiant radius, so that at first quadrature both stars would just fit onto a single spatial pixel.

The portions of the VV Cep spectra that are produced by an extended source are associated with the M-star's wind. Evidence is seen in the strong 1-eV Fe II and Mn II features in Fig. 3.7, where their intrinsic emission has been sculpted by absorption from a wind moving at $-20\,\mathrm{km\,s^{-1}}$ and thereby creating two apparent peaks. The longer-wavelength peak appeared to be associated with an extended source both in and out of eclipse and its intensity changed little, suggesting an origin in the extended regions of the M-supergiant wind, but during egress the shorter-wavelength one became increasingly augmented by emission from material that was being accreted onto the hot companion; 2-D spectra confirmed that it did not arise from a spatially-extended region. The total-eclipse observation plotted in Fig. 3.7 was in fact the last one obtained of that region during totality, and by then almost all of the flux in the short-wavelength component had emerged from eclipse.

Emission from the extended source revealed velocity structure, often with a change in sign that was correlated with the position angle of the detector, and was consistent with models of wind flow in a binary system in which the wind velocity is comparable to the orbital velocity.

3.11 A Simple Wind Density Model

The construction of an empirical model of the chromosphere and wind of the M supergiant from analyses of these complex spectra is a daunting challenge. In principle one can use information corresponding to different lines of sight through the extended atmosphere of the M star during chromospheric eclipse, and from that build an empirical model of the M-star's chromosphere and wind.

A useful starting point could be the damping wings of the strongest chromospheric absorption lines. Since the Lorentzian profile that gives rise to damping wings is so broad, the complex velocity structure that is present in the different lines of sight is unimportant—only the total line-of-sight column densities matter. Damping wings are only observed in a few of the strongest lines in the UV near

[1]For point-like sources, 2-D data are combined into 1-D spectra via standard pipeline processing.

eclipse: Lyα, Mg II h and k, and the Fe II resonance line at λ 2599 Å. Figure 3.11 fits a Lorentzian model to an observation of Mg II h and k during chromospheric eclipse.

Fitting the theoretical profile to that observation yields the total hydrogen column density, N_H, along the line of sight at that phase. Figure 3.12 plots $\log N_H$ as a

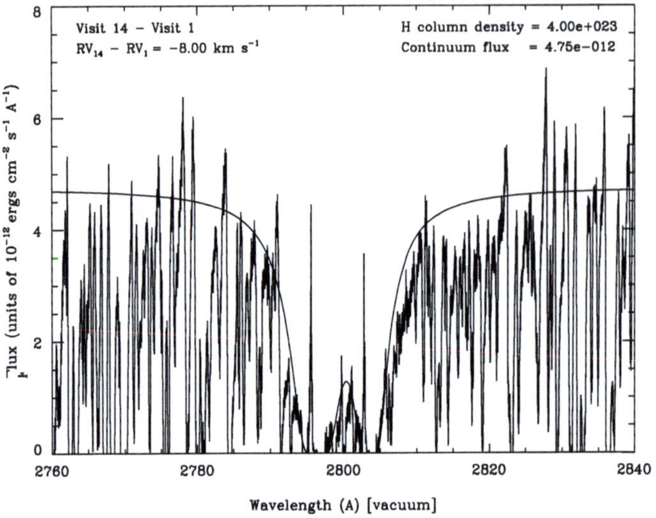

Fig. 3.11 Lorentzian model fitted to the damping-wing profiles of Mg II h and k as extracted from an observation at mid-chromospheric eclipse after subtracting the totality spectrum observed on 1997 Nov 29

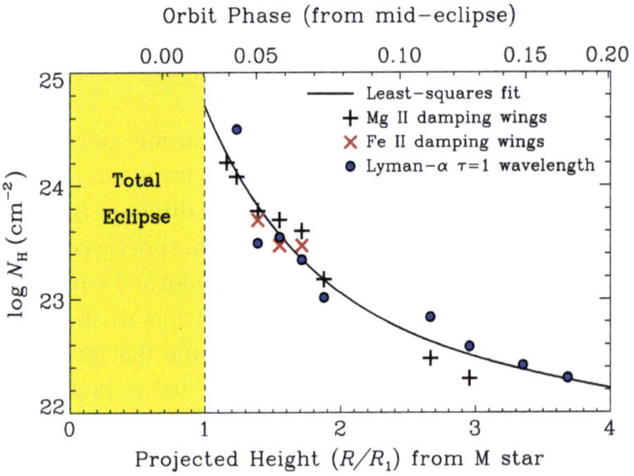

Fig. 3.12 Individual hydrogen column densities along the line of sight to the hot companion, calculated from the observed damping wings of Lyα, Mg II h and k, and Fe II λ 2599 Å, are plotted as a function orbital phase from mid-eclipse (top axis), and of projected radial distance of the line of sight to the hot companion (lower axis, with the stellar limb at $R/R_1 = 1$). The coloured/filled region represents total eclipse. The smooth curve is an empirically-derived least-squares fit

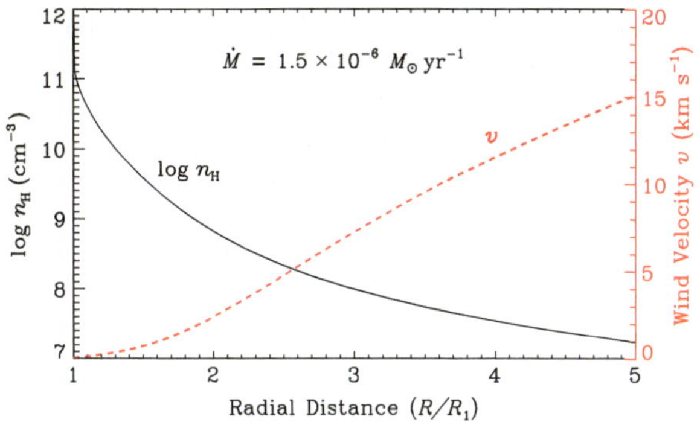

Fig. 3.13 *Solid line, left axis*: Empirical model of the hydrogen number density as a function of radial distance (with the stellar limb at $R/R_1 = 1$), derived from damping wings observed in *HST* spectra. *Dotted line, right axis:* Model of the wind velocity, as a function of the same radial distance, derived from that empirical model and assuming a steady spherical wind

function of phase from mid-eclipse, and of radial distance from the M-star's centre. Solar abundances, and a dominant single ionization stage for Mg and Fe, were assumed. Orbital phases were transformed onto a radial distance scale by adopting the line-of-sight impact parameters derived from Wright's (1977) orbit solution.

Under the assumption of a steady spherical flow, the fitting to the model can be inverted to recover the hydrogen *number* density as a function of radial distance. That has been done in Fig. 3.13, where the smooth line shows log N_H as a function of radial distance. An assumption of spherical symmetry is probably reasonable, but there is increasing evidence that these supergiant winds are episodic in nature rather than steady. Nevertheless, the model provides a reasonable time-averaged wind density, although it may need to be modified over decades. Despite these somewhat simplified assumptions, a velocity law can be derived; for that purpose the mass-loss rate is a free parameter, and the terminal velocity of the wind is proportional to the choice of $\dot{M}$. If we choose $\dot{M} = 1.5 \times 10^{-6}\, M_\odot$ year^{-1}, the corresponding terminal wind velocity is $\sim$20 km s^{-1}, and is represented by the dotted line in Fig. 3.13.

3.12 Looking Ahead

Analyses of extensive series of *IUE* and STIS UV spectra from two successive orbit cycles have enabled the complex nature of VV Cep to be at least partially understood. The high quality of the STIS spectra in both spectral resolution and S/N proved crucial, as also were the observations made during total eclipse.

The modelling summarized in this chapter represents first steps towards a complete semi-empirical model of the chromosphere and wind of this M-supergiant binary, but it has already raised questions. For example, there is little evidence of any wind acceleration regime in the *HST* spectra; prominent chromospheric features are present, as also are wind features shifted in velocity by about $-20\,\mathrm{km\,s^{-1}}$, but there is little absorption at intermediate velocities. The implication is that either the wind acceleration is rapid, so little column density was seen at intermediate velocities, or episodic, so no wind happened to be active at the times of those *HST* observations. In either case, the lack of much evidence of wind acceleration in the chromosphere itself suggests that the region of wind acceleration lies outside the extended chromosphere, or at a distance of at least 2–2.5 stellar radii from the centre of the star. Just how the wind is accelerated remains a mystery. However, radiative acceleration by dust grains is likely to be unimportant in this M supergiant because VV Cep shows very little evidence of dust compared to equivalent single stars such as α Ori, even though its wind and those from the single stars share properties in common.

Much work remains to be done to understand VV Cep and other eclipsing binaries better. We need to establish firmly the relevance—if any—of the system of VV Cep to other binaries such as KQ Pup, and especially to interpreting the chromospheres and winds of single red supergiants such as α Ori. Although *IUE* and *FUSE* are now long gone, and *HST* is nearing the end of its planned lifetime, those missions leave behind a rich archival legacy of ultraviolet observations, comprising a splendid resource for further examinations of what the spectra reveal, particularly if supported by more sophisticated modelling tools and data management software. It is our hope that further and deeper analyses will eventually answer questions of major importance, such as the causes and mechanisms of mass loss in late-K and M supergiants.

References

Bauer, W.H., Bennett, P.D.: PASP **112**, 31 (2000)
Bauer, W.H., Bennett., P.D., Brown, A.: ApJS **171**, 249 (2007)
Bauer, W.H., Gull, T.R., Bennett, P.D.: AJ **136**, 1312 (2008)
Bauer, W.H., Stencel, R.E., Neff, D.H.: A&AS **90**, 175 (1991)
Eggleton, P.P.: ApJ **268**, 368 (1983)
Faraggiana, R., Selvelli, P.L.: A&A **76**, L18 (1979)
Hack, M., Engin, S., Yilmaz, N.: A&A **225**, 143 (1989)
Hack, M., Engin, S., Yilmaz, N., et al.: A&AS **95**, 589 (1992)
Hagen, W., Black, J.H., Dupree, A.K., et al.: ApJ **238**, 203 (1980)
Hirata, R., Horaguchi, T.: Atomic Spectral Line List, SIMBAD On-line Data Catalog, VI/69 (1995)
Hutchings, J.B., Wright, K.O.: MNRAS **155**, 203 (1971)
Kawabata, S., Saitō, M.: PASJ **49**, 101 (1997)
Leedjärv, L., Graczyk, D., Mikolajewsi, M. et al.: A&A **349**, 511 (1999)
Möllenhoff, C., Schaifers, K.: A&A **94**, 333 (1981)
Pacyński, B.: ARA&A **9**, 183 (1971)
Savage, B.D, Sembach, K.R.: ApJ **379**, 245 (1991)
Stencel, R.E., Potter, D.E., Bauer, W.H.: PASP **105**, 45 (1993)
van de Kamp, P.: AJ **82**, 750 (1977)
Wright, K.O.: JRASC **71**, 152 (1977)

Chapter 4
ϵ Aurigae: A Two Century Long Dilemma Persists

Robert E. Stencel

Abstract The bright star ϵ Aurigae, an eclipsing binary with the longest known orbital period and eclipse, has been slow to reveal it secrets. The most recent eclipse, 2009-2011, was monitored with an array of new and old astronomical techniques, leading to new evidence concerning the nature of the eclipsing body and its contents. While this has led to new insights about the state of the F superigant star and to more clues about the evolutionary state of the system and components, a high-mass versus low-mass model dilemma persists. This chapter examines some of the recent developments and attempts to place the system among the pantheon of binary types. It also offers suggestions for observational tests prior to the next eclipse in 2037.

4.1 Introduction

ϵ Aurigae is arguably one of the most intriguing stars in the sky. First documented as a variable star by Fritsch in 1824, this object was declared an enigma as soon as binary-star analysis tools, refined by Russell (1912 a,b), were applied by Ludendorff (1906), Kuiper et al. (1937) and many others early in the twentieth century. The star is an F0 Ia supergiant that undergoes a 2-year-long, flat-bottomed eclipse every 27.1 years, the longest orbital period and eclipse of all known eclipsing binaries. The length of eclipse implies that the secondary component must be huge, but at minimum light the supergiant's photospheric spectrum still dominates. In the optical region, the eclipse is technically 'grey', being about 0.8 mag. deep at all wavelengths. No secondary eclipse has ever been recorded. These characteristics quickly led to the realization that the occulting body is not a normal star, and that the main source of opacity must be due almost entirely to scattering by electrons, aerosols or large grains. A geometrical model by Huang (1965), involving the passage of a thick, edge-on, opaque disk across the primary star, gained early favour and is the genesis of current, more physical models.

R.E. Stencel (✉)
Department of Physics & Astronomy, University of Denver Observatories, University of Denver, Denver, CO 80208, USA
e-mail: robert.stencel@du.edu

T.B. Ake, E. Griffin (eds.), *Giants of Eclipse: The ζ Aurigae Stars and Other Binary Systems*, Astrophysics and Space Science Library 408,
DOI 10.1007/978-3-319-09198-3_4

The latest eclipse occurred from 2009 to 2011, and was investigated world-wide by an assortment of instruments and observing techniques. As the interval since that event lengthens, so do opportunities for reflection on what the combined results are telling us. Many fundamental questions are still open: How far away is the system? What are the masses of its component stars? What is their evolutionary state? As observational tools have improved, each of the past five eclipses has yielded data against which theories can test their mettle. This chapter focuses on reports which largely resulted from the 2010 eclipse and augment the discussion by Stencel (2012). For a comprehensive historical review, the reader is referred to Guinan and DeWarf (2002) and to references therein.

4.2 Selected Results of the 2010 Eclipse Campaigns

4.2.1 *Orbital Solutions*

Two new orbital solutions, by Stefanik et al. (2010) and by Chadima et al. (2010), complement earlier ones by Morris (1962) and by Wright (1970). Those newer solutions largely agree, but have noteworthy differences; they are compared in Table 4.1, which lists Stefanik's 'combined fit' solution and Chadima's PHOEBE solution. Times of periastron have been phased by two and three orbital periods in the case of Stefanik and Morris, respectively, to facilitate comparison.

The constraint on the 89° inclination has been tightened as a result of in-eclipse interferometric imaging (Kloppenborg et al. 2010), which established that the disk crosses the southern hemisphere of the 2 mas-wide F-supergiant primary.[1] Owing to the high inclination, the angle of the line of nodes is less well constrained. One additional observational constraint comes from Portravnov's (2012) study of the

Table 4.1 Comparison of orbital element solutions

Element	Stefanik et al. (2010)	Chadima et al. (2010)	Morris (1962)
Period	9896.0 ± 1.6 days	9890.26 ± 0.6 days	9891 days
Eccentricity	0.227 ± 0.01	0.256 ± 0.01	0.17
ω	39°.2 ± 3°.4	41°.2 ± 3°.1	348°
$T_{\text{periastron}}$ (JD)	2 454 515 ± 80	2 454 596 ± 23	2 453 114
γ	-2.26 ± 0.15 km s^{-1}	—	-1.29 km s^{-1}
K	13.84 ± 0.23 km s^{-1}	14.40 ± 0.38 km s^{-1}	14.71 km s^{-1}
$f(M)/M_\odot$	2.51 ± 0.12	—	3.12
$a_1 \sin i$	12.2 ± 0.02 AU	—	13.1 AU

[1]The disk dimensions reported in that paper were based only on early eclipse phases, and should be used with caution.

F-star's axial rotational velocity, which was based on Fourier analysis of the cross-correlation maximum between observed and theoretical spectra and yielded a value of 38 km s^{-1} for that parameter.

4.2.2 *Spectral Energy Distribution*

Fundamental work of note includes studies of the spectral energy distribution (SED), particularly by Hoard et al. (2012), who assembled an SED from the far UV out to the far red at 250 μ. The SED shows that the spectrum is dominated by the flux from the high-luminosity F star, but it also indicates excess flux in the far-UV and in the infrared, which has been interpreted as arising from a hot star (of spectral type somewhere near B5 V) inside the disk, plus radiation from the disk itself. By comparing infrared photometry at primary conjunction with that at secondary conjunction, Stencel et al. (2011) demonstrated that the temperature of the disk ranges between 550 K on the side facing away from the 7500-K F star (i.e., during eclipse), to 1100 K on the side facing the F star (i.e., during secondary conjunction). Those differences were studied by Pearson and Stencel (2012) in an attempt to determine the characteristics of the material in the disk.

An attempt to detect the system in X-rays was reported by Wolk et al. (2010); they used the *XMM–Newton* spacecraft and a 25-ks integration, and determined a limiting X-ray flux of 2.5×10^{-15} erg cm^{-2} s^{-1}, suggesting a hydrogen column density $\sim 10^{23-25}$ cm^{-2}. That then provided a significant constraint on the density of the disk.

4.2.3 *Distance to the System*

Key to solving the component masses and the evolutionary status of the ϵ Aur system is determining the distance. An astrometric solution by Strand (1959) yielded a distance of 943 pc, while another by Van de Kamp (1978) gave 580 pc. The *Hipparcos* parallax has a large uncertainty, but it provides a distance of 650 pc and that value is adopted here for reference until a more precise determination has become possible. The apparent magnitude, $V = 3.05$, combined with a nominal reddening, *E(B–V)*, of 0.3 mag. and an extinction, A_V, of 3 *E(B–V)*, leads to an implied absolute magnitude, M_V, of -6.9 mag. If *E(B–V)* is a combination of interstellar and circumstellar reddening, as is indeed suggested by the variations in both photometry and colour indices, then M_V could be slightly lower than that but still be consistent with tabulated values for F supergiants like ϕ Cas (F0 Ia). The corresponding value of $\log L/L_\odot$ is $\sim$4.6.

The interferometric limb-darkened diameter of 2.27 ± 0.11 mas derived in the near-IR by Stencel et al. (2008) at the adopted reference distance of 650 pc implies a radius of 0.7 AU. or 160 $R_{\odot}$, which is a bit larger than for most Cepheids. If the internal structure and a 67-day pulsation (Sect. 4.2.4) resemble those of Cepheids, then the period-luminosity relationship (Benedict et al. 2007) would yield $M_V = -6.1$ mag. However, given the dispersion of derived masses for yellow supergiants, that association cannot be regarded as definitive.

Guinan et al. (2012) attempted to estimate the system's distance by comparing parallax, reddening and interstellar absorption of stars of B and A types within one degree of the line of sight to ϵ Aur. The correlations which they established between $E(B\text{–}V)$ and the equivalent width of the diffuse interstellar band at λ 6613 Å indicated that ϵ Aur is at a distance of 1.5 ± 0.5 kpc. Their use of a mean $E(B\text{–}V)$ colour overstates the interstellar contribution in this colour-variable star, so the interstellar line indicates an even larger distance. Combining that distance with the interferometric diameter of the F-star implies that the star has the huge radius of 1.7 AU (366 $R_{\odot}$). That combination of size and distance with an effective temperature of 7000 K corresponds to $\log L/L_{\odot} = 5.3$, and signals rapid evolution—but that may be inconsistent with the survival of a dense disk. It may be possible to repeat their experiment by using polarization of field stars as a check. However, in the meantime that result for the distance to ϵ Aur should be considered an outer limit to the likely reality.

Another indirect method to establish the distance to the system could involve using the disk parameters to constrain the binary parameters. Interferometric imaging (Kloppenborg et al. 2010, 2012) provides angular measures of disk length and thickness. The overall length is tidally (and thermally) bounded by the size of the semi-major axis, while the disk thickness is responsive to the gravitational potential of the central mass of the disk. Establishing either or both could help constrain the component masses and their evolutionary state. The duration of the near-infrared interferometric imaging that has been carried out (Kloppenborg 2012) corresponds to a disk length of 9.9 mas, which is consistent with the timing of ~700 days between first and fourth contacts (it is about 7% of the orbital period). Such imaging also shows that the most opaque portion of the disk eclipses the southern half of the F star but leaves a small region near the southern pole of the F star still visible, thereby constraining the angular thickness of the disk to ~0.8 mas. At the reference distance of 650 pc, those angular sizes correspond to a disk length of 6.4 AU and a thickness of 0.5 AU, as seen in the near-infrared. Spectro-interferometry of Hα (Mourard et al. 2012) indicates that the thickness of the gas disk is greater than the diameter of the F star, thus illustrating the response of lighter materials to the central gravitational potential, and indicating a mass for its central star of 6 $M_{\odot}$.

Yet another approach to constrain the system's distance involves using the dust properties of the disk to argue that the local effective temperature in the outer edge of the disk facing the F star is below the level for sublimation, and thus to establish a minimum separation between the disk's edge and the F star (Pearson and Stencel 2012). For the range of widely acceptable distances to the system, given that the separation and F-star diameter scale accordingly, any differential heating assumes that the UV output of the F star be mass-dependent, and that the local heating depends on the disk's mass and on the opacity and sublimation properties of the dust. For example, if the system is at or beyond one 1 kpc in distance, the implied high luminosity of the F star could be too great to allow solids to exist in the disk if it has a relatively low mass. In such a case, it would be worth looking for any unique spectral signatures that high-density aerosols might show.

4.2.4 Light Curves

Worldwide interest in the 2009–11 eclipse of ϵ Aur yielded the most comprehensive set of light curves for any of its eclipses to date; they span the entire optical region (*UBVRIJH*), thanks largely to campaigns organized by the AAVSO (Percy 2012) and by Hopkins (2012). The new light curves are largely consistent with those of prior eclipses, but show more detail concerning in-eclipse variations and a putative mid-eclipse brightening (Fig. 4.1). The in-eclipse variations with a quasi-periodic 67-day cycle are most clearly discerned in the *U*-band light curve. Furthermore, during this recent eclipse the system was faintest in the *U*-band

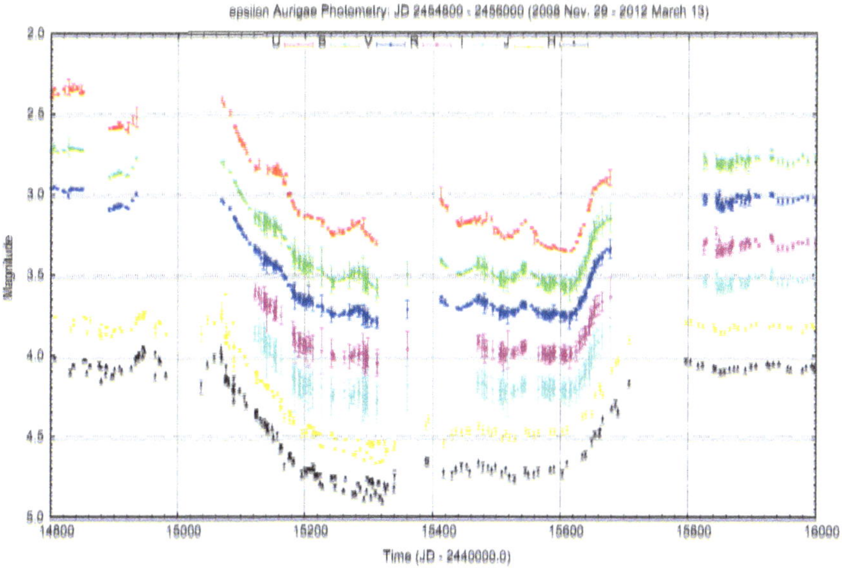

Fig. 4.1 *UBVRIJH* light curves of the 2009–2011 eclipse of ϵ Aur ranged downwards (*Reproduced, with kind permission, from Kloppenborg et al. (2012)*)

just prior to third contact, but less so in the R- and I-bands. Ingress proceeded more slowly than egress, with the exception of a 'still-stand' occurring between third and fourth contacts, which may be associated with a mass exchange stream discovered spectroscopically (Griffin and Stencel 2013). Attempts to detect shorter-term variations of hours or minutes have not found any consistent signal, perhaps thereby ruling out a compact source in the disk.

The existence of variability of $\sim$0.2 mag. both in and out of eclipse is one of the long-known features of the ϵ Aur light curve. Authors have associated those variations with instabilities in the F star, adducing either pulsations or spots. Combining the 67-day quasi-period of those variations with the rotational velocity of $38\,\text{km}\,\text{s}^{-1}$ (Sect. 4.2.1) and a radius of $160\,R_{\odot}$ implies a rotation period of the star of 212 days, which is roughly three times the quasi-period.

One of the most comprehensive analyses of the photometric variations was that by Kloppenborg et al. (2012), who analyzed virtually all available *UBVRIJH* photometry and discovered that the F-star variations are multi-periodic with at least two periods that evolve in time and at a rate of nearly -1.5 day per year. The periods are best detected when they manifest themselves as near-sinusoidal variations at 3200-day intervals. This conjecture is testable by continued photometric monitoring.

Wilson et al. (2013) combined photometry and spectroscopy and applied the Baade-Wesselink method to estimate the distance to the system. Their method is promising, but should be applied when the variations are coherent in order to avoid other misleading complexities inherent in the radial-velocity (RV) signal.

In their analyses of the photometric and spectroscopic variations of ϵ Aur, Harmanec et al. (2013) suggested the occurrence of episodes of atmospheric mass transfer from the F star towards its companion; those might be consistent with the mass-transfer stream detected by Griffin and Stencel (2013). Kambe et al. (2013) reported period analyses of the variations seen in the F-star's photospheric lines, indicating that there are multiple periodicities similar to those seen in the photometry. The excitation of non-radial pulsations in the primary was considered as a possible explanation for the dominant period of about 67 days.

4.2.5 Spectroscopy

Since the opaque disk in ϵ Aur does not have a detectable spectrum of its own, information about its temperature, density and sub-structure has had to be prised out of the signatures which it adds to the observed spectrum of the system at eclipse phases, particularly during ingress and egress. In that respect ϵ Aur certainly resembles the ζ Aur binaries (see Fig. 4.3), but although it has often been associated with that group in the past, it has mostly been on account of the many puzzling differences rather than through any apparent (but misconceived) physical parallels.

During recent years telescopes large and small, equipped with spectrometers, have contributed numerous spectra at eclipse phases. Spectroscopy is key to the latest refinements of the orbital solutions by Stefanik et al. (2010) and by Chadima et al. (2010), as well as to the discoveries and enhancements that have resulted from re-examination of prior efforts.

Reports of heroic levels of spectroscopic monitoring include—but are not limited to—outstanding Hα spectral histories spanning a number of years both before and after eclipse. The reports include papers by Chadima et al. (2011), Mauclaire et al. (2012), and Griffin and Stencel (2013), as well as work in progress by Strassmeier with robotic equipment (Granzer et al. 2010), and a first application of spectro-interferometry by Mourard et al. (2012). Those observations followed the development of the eclipse of the F star by the disc in Hα, Na D, and various metal lines. Efforts to disentangle the spectral components in ϵ Aur have been reported by Chadima et al. (2012), who found that the line-profile variations which were seen were not consistent with the presence of a secondary star. A similar attempt, reported by Bennett et al. (2012), to detect evidence of the secondary component by examining the centroids of prominent absorption lines in the red spectral region in the hope of measuring centroid shifts that could be attributable to RV shifts of the secondary component was thwarted by line-profile variations and asymmetries that were intrinsic to the F-star primary. However, both efforts assumed that a pair of absorption lines would contribute adequately to the observed signal, whereas the indications are that, in addition to the F-star's photospheric absorption, there may also be scattered light from an extended nebulosity within the system. Re-analysis with that more general approach might reveal new information.

A particularly thorough spectroscopic investigation was the comprehensive study of neutral potassium by Leadbeater et al. (2012), parallelling the discovery by Parthasarathy (1982) and the eclipse coverage through that line by Lambert and Sawyer (1986). Outside eclipse, the λ 7699 Å K I line simply appeared as a weak interstellar feature, whereas during eclipse it strengthened quite impressively and exhibited velocity signatures of the disk in considerable detail. The excess absorption started to grow after first contact, and reached a maximum near second contact. It then declined slightly through mid-eclipse, only to rebound and grow even larger until third contact, and dwindled back to the weak interstellar level at and after fourth contact. Rather than the pattern of line-strength variations being attributable to an actual abundance effect, they seem to describe structure within the disk, possibly as a result of fractional ionization brought about by radiative stripping of low-ionization elements in the disk—a phenomenon seen on airless worlds of our solar system, such as Mercury and Io (Schaefer and Fegley 2009).

Another striking result of the intensive monitoring by Leadbeater et al. is the discovery of discontinuous step-like changes in the strength of the potassium absorption. Leadbeater and Stencel (2010) interpreted those steps in terms of ring-like structures within the disk, possibly related to tidally-driven spiral density waves. Since periastron slightly precedes the eclipse phase, it could be supposed that an impulse might be imparted to the disk. If the disk rotation rate is faster than, and not synchronized to, the orbital rate (Griffin and Stencel 2013), then the disk may be dynamically young and definitely not in equilibrium—although quasi-equilibrium or resonance is possible.

High-S/N spectroscopy obtained during the most recent eclipse yielded new evidence about the sub-structure of the disk. Griffin and Stencel (2012, 2013) and Sadakane et al. (2013) both reported spectroscopic features that were consistent with a mass-exchange stream on the trailing edge of the disk. Griffin and Stencel (2013) additionally reported the transient appearance of rare-earth elements during third contact when stream features were strongest (see Fig. 4.2). Interpreted in the context of a wind-capture mass-loss, the stream density could replenish the disk against central accretion at a rate that was consistent with the multi-million year evolutionary timescale of the F star.

As mentioned above, the 'shell' lines, which are the only easily-detected spectroscopic signatures of the ϵ Aur disk, closely resemble those of a cool-star chromosphere. Griffin and Stencel (2013, their fig. 12) drew attention to that close similarity by direct comparison. They compared the shell spectrum of ϵ Aur (which they derived by subtracting, from an egress spectrum, an observation made outside eclipse and during an apparent lull in the frequent episodic line-profile changes) with the spectrum of the chromosphere of ζ Aur at an egress phase shortly following fourth contact; the latter had been derived by subtracting a total-eclipse spectrum from the egress one. The close similarity (illustrated here in Fig. 4.3) is pronounced, and suggests quite comparable temperatures, though the slightly narrower lines in the ϵ Aur spectrum also suggest lower pressure.

Another comparison with ζ Aur stars was drawn by Ake (2006), who discussed the rich emission-line spectrum of ϵ Aur seen in the far UV. Figure 4.4 illustrates a striking similarity between *FUSE* spectra of ϵ Aur and 22 Vul. Since those

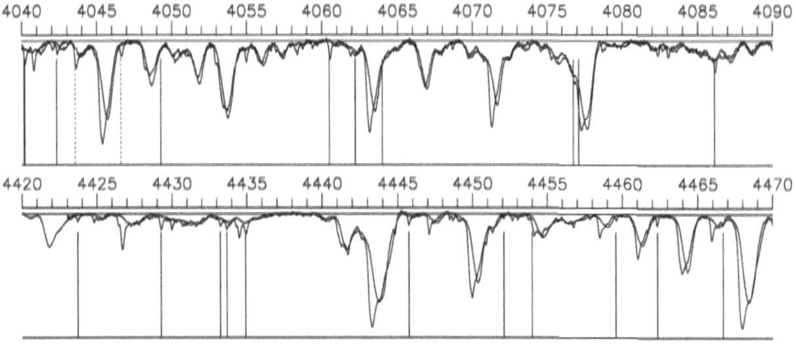

Fig. 4.2 Evidence of a transient mass-transfer stream. *Upper:* spectra at phases .117 (*thin line*) and .042 (*thick line*) in the region of Hδ. *Lower:* comparisons of spectra in the blue, observed at phases .116 (*thin line*) and .022 (*thick line*). *Solid vertical lines* mark the positions of rare-earth lines, blue-shifted by 40.5 km s^{-1}. The (*dotted lines*) in the upper panel correspond to ground state transitions of K I. (Data source: DAO CCD spectra recorded by Griffin)

emission lines can definitely be attributed to scattering photons in the hot star, Ake suggested that a parallel interpretation is also possible in the case of ϵ Aur, proposing the source(s) as resonance scattering of photons coming either from the expanding wind of the supergiant, or from an occulted hot source inside the disk, or both.

Near-infrared spectroscopy reported by Stencel et al. (2011) included the discovery of strengthening of He I λ 10830 Å during mid-eclipse, along with a recurrence of transient molecular CO during the second half of eclipse (Stencel and Geballe 2014; Hinkle and Simon 1987). That He I line is an indicator of an ionizing source in the disk centre, and appears consistent with what would be seen from a mid-B main-sequence star. The lack of minerals among the mid-infrared spectroscopic features is taken to imply that the disk contains large grains, although their composition and specific sizes have not been determined (Budaj 2011, 2012; Muthumariappan and Parthasarathy 2012).

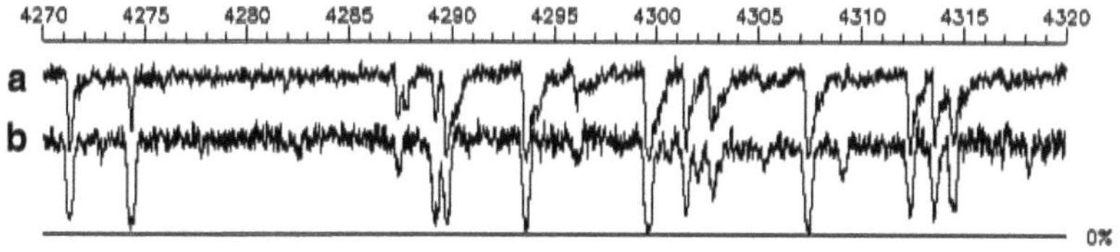

Fig. 4.3 **a**: A spectrum of the ϵ Aur disk, extracted from a Mt.Wilson observation at phase .121 of the 1956–1957 eclipse, and shifted in wavelength to match **b**, a spectrum of the chromosphere of ζ Aur, extracted from a spectrum of the system observed at Calar Alto by Griffin during its 1990 eclipse. The zero line applies to the spectrum in **b**; the one in **a** has the same vertical scale as that in **b** but has been raised for clarity *(Reproduced, with kind permission, from Griffin and Stencel (2013))*

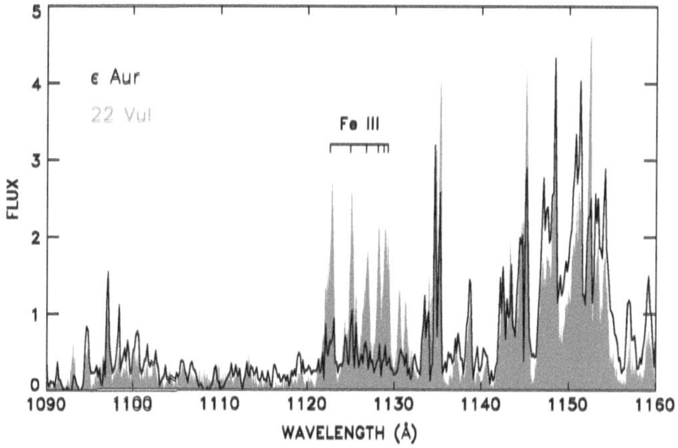

Fig. 4.4 *FUSE* (far-UV) spectrum of ϵ Aur (*black*) compared to 22 Vul (*grey*). The prominent lines in the spectral region λ 1142–1154 Å arise from Fe II; as Fig. 2.26 demonstrates, in 22 Vul they are a recognized signature of wind scattering

A minor but potentially significant discovery prior to the recent eclipse was the detection of a large blue-shifted feature in Hα. It occurred before, but close to, periastron, and was observed serendipitously in 2006 by both Harrington and Kuhn (2009; their fig. 33a) and by Chadima et al. (2011; their fig. 2), and was coincident with the first Spitzer infrared observations of the system (Stencel 2007). One suggestion is that tidal effects, augmented by the slightly closer approach, triggered an event of mass ejection from the F star. Observational confirmation during next periastron (*circa* 2033) and approach to the next eclipse is clearly to be encouraged.

4.2.6 Additional Results, and New Work in Progress

Important new results have emerged in the areas of polarimetry, spectro-polarimetry and spectro-interferometry. The series of broad-band polarimetric observations reported by Cole (2012) confirm and extend results by Kemp et al. (1986), including a pronounced maximum in polarization at third contact. A series of 50 spectro-polarimetric observations, obtained with the ESPaDOnS high-resolution instrument on the *CFHT* and bracketing the 2009–2011 eclipse, have been analysed by Geise et al. (2012). The data reveal strong, wavelength-dependent polarization changes in numerous spectral features (see also Harrington and Kuhn 2009). Spectrally-resolved optical interferometry, obtained with the VEGA beam combiner on the CHARA array and spanning the 2009–2011 eclipse, showed that the disk is larger when seen in the core of Hα (and possibly eclipsing the entire F star) compared to the near-infrared, where only one hemisphere seems to be obscured during eclipse Mourard et al. (2012). The scale height of the gas in the disk may therefore be much larger than the scale height of the dust, thereby indicating that the unseen star at its centre can be less massive than is required by the high-mass models.

Automated spectroscopy is also very appropriate for such a bright star. In that mode (Eaton and Williamson 2007), Eaton obtained 601 high-resolution spectra (R $\sim$ 30,000) covering λ 5000–7100 Å, for a span of nearly 5 years centred on the recent eclipse.

4.2.7 Relative Component Masses: Models

The history of the study of ϵ Aur includes several attempts to determine the component masses. The high luminosity and low surface gravity of the F star originally suggested that it was a massive supergiant star. However, as the study of stellar evolution has progressed, a whole class of lower-mass, high-luminosity post-AGB stars has been recognized (see, for example, Hajduk et al. 2008). Those, too, bear a resemblance to the F star in ϵ Aur (allowing for details of differing abundances). If it can be assumed that the mass function is ~2.5 (Table 4.1), those solutions that yield a mass ratio q (defined here as $M_{\text{Fstar}} : M_{\text{companion}}$) that is close to unity imply high masses, and those with q close to 0.5 imply lower masses. Most solutions necessarily include a series of assumptions, and critical tests of each assumption are needed.

The low-mass models include those by Handbury and Williams (1977), Lambert and Sawyer (1986), Takeuti (1986), Saito et al. (1987) and Hinkle and Simon (1987). Proponents of high-mass models tackled the problem by re-interpreting the uncertain distance and reddening of the system through various routes. Two hybrid models, proposed by Lissauer and Backman (1984) and by Eggleton and Pringle (1985), hide the companion's luminosity by invoking a triple-system solution. In an effort to resolve the degeneracy, Stencel and Geballe (2014) observed the post-eclipse velocity offset between the F star's photospheric infrared Huθ and nearby Brα emission, in order to test whether the latter arises from the disk centre, which is observed to be the source of strong He I λ 10830 Å absorption. However, they found a velocity splitting only half of that expected from the high-mass orbital solution, implying either that the disk is the more massive, or that the emission arises from an extended region nearer the system's barycentre. Further observations are clearly needed. It should be mentioned that the Brα emission line was observed before the 2009–2011 eclipse by Clemens et al. (2007, their fig. 15), and during the 1984 one by Backman et al. (1985).

4.3 Concordances with Other Binary Systems

A number of papers make mention of ϵ Aur on account of presumed similarities in observational properties. The system EE Cep is one such; it is a 5.6-year eclipsing binary containing a Be star and an invisible companion surrounded by a dark, dusty disk. According to comprehensive photometry reported by Gałan et al. (2012), the three prior eclipse cycles were showing decreasing depths of totality. Moreover, multiple components are seen to appear in spectral lines during eclipse. Gałan et al. concluded that the disk is highly structured, that the B star has polar hot spots which contribute to the light curve, that the disk precesses, and predicted that the 2014 eclipse should be deeper than earlier ones. Particular similarities with ϵ Aur include evidence for disk sub-structure, and that hot spots on the primary star may affect the overall light curve. Although at $V = 11$ mag. EE Cep is considerably fainter than ϵ Aur, high-resolution studies are within reach with large telescopes.

BM Ori, the faintest of the four stars located within the Trapezium, is a 6.47-day binary containing a B3 star that is eclipsed by a mysterious secondary. Windemuth et al. (2013) report that *MOST* photometry through 4 orbits shows a deep, symmetric primary eclipse with small cycle-to-cycle variations. Furthermore, in comparison to light curves from a decade ago, the recent ones have become more shallow and totality has become shorter. They model the secondary as a self-luminous, reflective disk surrounding an extremely young protostar, and had the disk undergoing opacity evolution during the past century. Their model of this young, compact, interacting binary also includes leading and trailing streams of material. BM Ori can be compared to ϵ Aur in that tidal forces and disk-irradiation play strong roles in the observed changes.

V582 Mon (KH15D), whose properties have been extracted by Windemuth and Herbst (2014) from decades of photometry, is a 48.37-day eclipsing system comprised of an eccentric T Tauri binary embedded in an inclined and precessing circumbinary ring. The brightness of the system has been increasing as the B star emerges from the trailing edge of the occulting material. Colour variations suggest that the companion is a similar type of B star, and that the circumbinary material consists of large dust particles. V582 Mon shares similarities with ϵ Aur in the notion of a circumbinary ring of material associated with the binary interaction, as indicated by polarization changes seen over many orbital phases.

M2-29, which is understood to be the first eclipsing-disk system detected among post-AGB stars, has exhibited an occultation event (Hajduk et al. 2008). The duration of the eclipse was nearly 3 years, and could be modelled with a binary period of 18 years. Although the disk in this case is circumbinary, the scale of the system suggests an analogy with ϵ Aur that may help in identifying the properties of the disk.

The system RW Aur presents a particularly dynamic case, interpreted as showing recent light-curve variations associated with a 'flyby' of its companion, RW Aur B, that caused tidal stripping of its circumstellar disk, leaving a spiral arm of material stretching towards it and RW Aur B (Rodriguez et al. 2013). When and whether we will see this phenomenon recur depends on observer vigilance. The relevance to ϵ Aur is that, with its relatively large eccentricity and a periastron that slightly precedes eclipse, tidal disruption of the disk is suspected—and may have been detected during the 2006 spectroscopic observations.

Another related example is the system KIC 12557548b. It was discussed by Budaj (2013) as a transiting but disintegrating extrasolar planet with a comet-like tail. Features in the light curve appear to be analogous to those in the ϵ Aur system, and are consistent with strong forward scattering from dust particles at eclipse phases. Polarization studies of ϵ Aur suggest that polarization effects should be looked for in KIC 12557548b.

OGLE-LMC-ECL-11893 is an eclipsing system with a 468-day period, and shows a deep, asymmetric eclipse that has repeated exactly in over 17 years of monitoring (Dong et al. 2014). Ingress and egress are steep, and the second half of eclipse is 0.5 mag. deeper in the I-band than during first half of eclipse—analogous to, but more extreme than, ϵ Aur during the 2009–2011 eclipse. The duration of eclipses is 15 days, or 3% of the binary period, and is thus about half the percentage in the case of ϵ Aur. The is aso the possibility of a tidally-limited circumstellar disk and stream. However, this binary and a similar one in the LMC suggest that the spatial frequency of these young, edge-on disks as in ϵ Aur are relatively rare (though large-scale surveys may incrementally modify those statistics).

V383 Sco is a system with a rather long period (13.5 years). Examination of long-term photometry and new high-resolution spectra led Gałan et al. (2013) to determine that the system is comprised of a pulsating M supergiant which eclipses an obscured but high-luminosity F0 star. Eclipses are deep, and vary in both duration and shape. It is tempting to consider V383 Sco as a slightly older version of ϵ Aur, in which the F star is in a second crossing and the upper main-sequence star has reached the red-supergiant phase. If that model is correct, it could be profitable to search for evidence of prior mass loss and circumstellar shells in far-infrared maps, although the low galactic latitude of V383 Sco means a complex line of sight.

The case of V383 Sco invites an examination of any connection between the evolutionary state of ϵ Aur and those of other ζ Aur systems. Studies of atmospheric-eclipse phases of the cooler ζ Aur systems have identified interaction shocks as the hot star ploughs through the red supergiant's extended wind; application of those ideas to ϵ Aur include identifying atmospheric structure in the F star, and searching for wind-interaction zones in the binary.

One working hypothesis for the evolutionary status of the ϵ Aur system has the F star recently evolved from the main sequence and on its first crossing of the upper H–R diagram. Interior changes will have produced a carbon dredge-up and wind enhancement event, capable of creating the ^{13}C enriched (Hinkle and Simon 1987) wind-capture disk mentioned in Sect. 4.2.5 and the rare-earth enriched stream reported by Griffin and Stencel (2013). A high-luminosity F star in its first ascent and crossing will not yet have undergone heavy mass loss of the sort needed to create the extended infrared shells that are typically associated with red supergiants (Stencel et al. 1989); indeed, inspection of far-infrared maps obtained with *IRAS*, *AKARI* and *Herschel* have not detected mass-loss shells that could be associated with ϵ Aur (Hoard et al. 2012). An abundance analysis by Sadakane et al. (2010) has provided support for that interpretation; they found that, while ϵ Aur does not show abundances indicative of strong post-AGB s-processes, there were slight enhancements that could be indicative of s-process nucleosynthesis in the stellar interior and dredge-up at some point in the star's history.

4.4 Key Observational Opportunities

The orbital phase of elongation, which occurs around 2016, can provide opportunities to resolve the disk from the central star by infrared methods. The orbit deduced by Stefanik et al. (2010) indicates that the greatest velocity separation, of more than 30 km s^{-1} between F star and disk, will last for a couple of years. Velocity cross-over (i.e. secondary conjunction) will occur during 2026–2027, when infrared techniques might again be fruitful since the side of the disk heated by the F star will be facing towards the Earth.

As the time to the next eclipse (in 2037) decreases, opportunities to prepare the next set of key observations gather momentum. Mid-eclipse, predicted for JD 2 465 290, occurs during 2037 August, which is somewhat farther from solar conjunction than was the case in 2010. Much more attention should be paid to third contact, using all available spectroscopic, polarimetric *and* interferometric methods to study the interesting egress phenomena that were detected in 2011 during that phase.

Since the F-star contribution at longer wavelengths is less than that of the disk, observations at sub-mm and mm wavelengths to detect molecules such as CO, CO_2 and H_2O in the disk would provide good bases for modelling the disk's composition. Those molecules are key to deducing the mass of the disk and its temperature structure, and can in turn strongly constrain the range of solutions for the mass and distance of the system as a whole. They will be challenging observations to make, but the 2009–2011 eclipse has helped us to focus on key aspects both in and out of eclipse, and has thus brought us closer to resolving questions of the distance and evolutionary status of the components that make up the ϵ Aur system.

References

Ake, T.B., in Astrophysics in the Far Ultraviolet, eds. G.Sonneborn, H.Moos, B-G Andersson (ASPC **348**, San Francisco, 2006), p.156

Backman, D.E., Simon, T., Hinkle, K.H.: PASP **97**, 1163 (1985)

Benedict, G.F., McArthur, B.E., Feast, M.W., et al.: AJ **133**, 1810 (2007)

Bennett, P.D., Harmanec, P., Chadima, P., et al.: in From Interacting Binaries to Exoplanets: Essential Modeling Tools, ed. by M.T. Richards, I. Hubeney (IAUS **282**, Cambridge, 2012), p. 399

Budaj, J.: A&A **532**, L12 (2011)

Budaj, J., in From Interacting Binaries to Exoplanets: Essential Modeling Tools, ed. by M.T. Richards, I. Hubeney (IAUS **282**, Cambridge, 2012), p. 293

Budaj, J.: A&A **557A**, 72 (2013)

Chadima, P., Harmanec, P., Yang, S., et al.: IBVS **5937** (2010)

Chadima, P., Harmanec, P., Bennett, P.D., et al.: A&A **530A**, 146 (2011)

Chadima, P., Harmanec, P., Bennett, P.D., et al.: in From Interacting Binaries to Exoplanets: Essential Modeling Tools, ed. by M.T. Richards, I. Hubeney (IAUS **282**, Cambridge, 2012), p. 401

Clemens, D.P., Sarcia, D., Grabau, A., et al.: PASP **119**, 1385 (2007)

Cole, G.M., JAVSO **40**, 787 (2012)

Dong, S., Katz, B., Prieto, J., et al.: arXiv1401.1195 (2014)

Eaton, J.A., Williamson, M.H.: PASP **119**, 886 (2007)

Eggleton, P.P., Pringle, J.E.: ApJ **288**, 275 (1985)

Gałan, C., Mikołajewski, M., Tomov, T., et al.: A&A **544A**, 53 (2012)

Gałan, C., Tomov, T., Kato, T., et al.: A&A **550A**, 93 (2013)

Geise, K., Stencel, R.E., Manset, N., et al.: JAVSO **40**, 767 (2012)

Granzer, T., Weber, M., Strassmeier, K.G.: Ad. Astron. **2010**, 1 (2010)

Griffin, R.E., Stencel, R.E.: JAVSO **40**, 714 (2012)

Griffin, R.E., Stencel, R.E.: PASP **125**, 775 (2013)

Guinan, E.F., DeWarf, L.E.: in Exotic Stars as Challenges to Evolution, ed. by C. Tout, W. Van Hamme, (IAUC 187, ASPC **279**, San Francisco, 2002), p. 121

Guinan, E., Mayer, P., Harmanec, P., et al.: A&A **546A**, 123 (2012)
Hajduk, M., Zijlstra, A.A., Gesicki, K., A&A **490**, L7 (2008)
Handbury, M.J., Williams, I.P.: Observatory **97**, 73 (1977)
Harmanec, P., Božić, H., Korčáková, D., et al.: C. Eur. Ap. Bull. **37**, 99 (2013)
Harrington, D.M., Kuhn, J.R.: ApJS **180**, 138 (2009)
Hinkle, K.H., Simon, T.: ApJ **315**, 296 (1987)
Hoard, D.W., Ladjal, D., Stencel, R.E., et al.: ApJ **748**, L28 (2012)
Hopkins, J.L.: JAVSO **40**, 633 (2012)
Huang, S.-S.: ApJ **141**, 976 (1965)
Kambe, E., Sadakane, K., Hashimoto, O., et al.: in Progress in Physics of the Sun and Stars: A New Era in Helio- and Asteroseismology, ed. by H. Shibahashi, A.E. Lynas-Gray (ASPC **479**, San Francisco, 2013), p. 203
Kemp, J.C., Henson, G.D., Kraus, D.J., et al.: ApJ **300**, L11 (1986)
Kloppenborg, B., Stencel, R., Monnier, J.D., et al.: Nature **464**, 870 (2010)
Kloppenborg, B.K.: Ph.D. Thesis (Univ. of Denver, 2012)
Kloppenborg, B.K., Hopkins, J.L., Stencel, R.E., JAVSO **40**, 647 (2012)
Kuiper, G.P., Struve, O., Strömgren, B.: ApJ **86**, 570 (1937)
Lambert, D.L., Sawyer, S.R.: PASP **98**, 389 (1986)
Leadbeater, R., Stencel, R.: arXiv 1003.3617 (2010)
Leadbeater, R., Buil, C., Garrel, T., et al.: JAVSO **40**, 729 (2012)
Lissauer, J.J., Backman, D.E.: ApJ **286**, L39 (1984)
Ludendorff, H.: AN **171**, 49 (1906)
Mauclaire, B., Buil, C., Garrel, T., et al.: JAVSO **40**, 718 (2012)
Morris, S.C.: JRASC **56**, 210 (1962)
Mourard, D., Harmanec, P., Stencel, R., et al.: A&A **544A**, 91 (2012)
Muthumariappan, C., Parthasarathy, M.: MNRAS **423**, 2075 (2012)
Parthasarathy, M., in ϵ Aurigae Newsletter **5**, 5 (1982)
Pearson, R.L., Stencel, R.E.: JAVSO **40**, 802 (2012)
Percy, J.R.: JAVSO **40**, 230 (2012)
Portravnov, I.S.: Ap. **55**, 528 (2012)
Rodriguez, J.E., Pepper, J., Stassun, K.G., et al.: AJ **146**, 112 (2013)
Russell, H.N.: ApJ **35**, 315 (1912a)
Russell, H.N.: ApJ **36**, 54 (1912b)
Sadakane, K., Kambe, E., Sato, B., Honda, S., Hashimoto, O.: PASJ **62**, 1381 (2010)
Sadakane, K., Kambe, E., Hashimoto, O., et al.: PASJ **65**, 1 (2013)
Saito, M., Kawabata, S., Saijo, K., Sato, H.: PASJ **39**, 135 (1987)
Schaefer, L., Fegley, B.: ApJ **703**, L113 (2009)
Stefanik, R.P., Torres, G., Lovegrove, J., et al.: AJ **139**, 1254 (2010)
Stencel, R.E.: in Binary Stars as Critical Tools and Tests in Contemporary Astrophysics, ed. by W.I. Hartkopf, E.F. Guinan, P. Harmanec (IAUS 240, Cambridge, 2007), p. 202
Stencel, R.E.: JAVSO **40**, 618 (2012)
Stencel, R.E., Pesce, J., Hagen-Bauer, W.: AJ **97**, 1120 (1989)
Stencel, R.E., Creech-Eakman, M., Hart, A., et al.: ApJ **689**, L137 (2008)
Stencel, R.E., Kloppenborg, B.K., Wall, R.E., et al.: AJ **142**, 174 (2011)
Stencel, R.E., Geballe, T.R.: AJ (2014, submitted)
Strand, K.Aa.: AJ **64**, 346 (1959)
Takeuti, M.: Ap&SS **120**, 1 (1986)
Van de Kamp, P.: AJ **83**, 975 (1978)
Wilson, C.K., Bennett, P.D., Hopkins, J.L., et al.: C. Eur. Ap. Bull., **37**, 109 (2013)
Windemuth, D., Herbst, W.: AJ **147**, 9 (2014)
Windemuth, D., Herbst, W., Tingle, E., et al.: ApJ **768**, 67 (2013)
Wolk, SJ., Pillitteri, I., Guinan, E., et al.: AJ **140**, 595 (2010)
Wright, K.O.: VA **12**, 147 (1970)

Chapter 5
Atmospheric Heating and Wind Acceleration in Cool Evolved Stars

Vladimir S. Airapetian and Manfred Cuntz

Abstract A chromosphere is a universal property of stars of spectral type later than ~F5. Evolved K and M giants and supergiants, including the ζ Aur binaries, show extended and highly turbulent chromospheres, which develop into slow massive winds. The associated continuous mass loss has a significant impact on stellar evolution, and thence on the chemical evolution of galaxies. Yet despite the fundamental importance of those winds in astrophysics, the question of their origin(s) remains unsolved. What mechanisms heat a chromosphere? What is the role of the chromosphere in the formation of stellar winds? This chapter provides a review of the observational requirements and theoretical approaches for modelling chromospheric heating and the acceleration of winds in single cool, evolved stars and in eclipsing binary stars, including physical models that have recently been proposed. It describes the successes that have been achieved so far by invoking acoustic and MHD waves to provide a physical description of plasma heating and wind acceleration, and discusses the challenges that still remain.

5.1 Introduction

Stars of spectral types later than about F5, including the Sun, possess convective zones, chromospheres and winds. The strong convective motions in giant and supergiant stars—thus also encompassing the primaries of ζ Aur systems—carry processed material, produced by nuclear-burning reactions, from the stellar interior to the atmosphere, where massive winds originate, and from there it is injected into the interstellar medium. Continuous mass loss therefore has a significant impact upon patterns of stellar evolution, especially in the late phases of a star's life,

V.S. Airapetian (✉)
NASA/Goddard Space Flight Center, 8800 Greenbelt Rd., Greenbelt, MD 20771, USA
e-mail: vladimir.airapetian@nasa.gov

M. Cuntz
Department of Physics, University of Texas at Arlington, 502 Yates St., Arlington, TX 76019, USA
e-mail: cuntz@uta.edu

T.B. Ake, E. Griffin (eds.), *Giants of Eclipse: The ζ Aurigae Stars and Other Binary Systems*, Astrophysics and Space Science Library 408,
DOI 10.1007/978-3-319-09198-3_5

upon the chemical evolution of galaxies (including the mass and energy budgets of the interstellar medium), and even upon the long-term evolution of exoplanetary atmospheres. Mass loss from stars can also affect planetary habitability through the dynamical and collisional evolution of planetesimals. By studying the physical mechanisms that drive those outflows, and also their interactions with stellar convection, rotation, pulsation and magnetic fields, we are better able to assess and quantify the importance of stellar winds for astrophysics in general. Hence, understanding stellar winds is one of the fundamental challenges in theoretical astrophysics. Clues to the nature of the acceleration of stellar winds should originate from their underlying base. In most of the cases (except for a small number of rapidly rotating giants such as YY Men), that base is represented by a chromosphere and/or a transition region that is heated to a few 10^5 K, while cooler giants and supergiants, say $(V - R) > 0.8$, manifest winds that emanate from the extended outer chromosphere.

The nett radiative energy flux from the chromosphere—the region above the photosphere—is more than ten times greater than that from the entire overlying transition region and wind. Chromospheric temperatures are too high to be explained by radiative heating alone, so there must therefore be some type of mechanical energy input. The convective zones constitute the major power source for such an energy flux; they contribute to the generation of UV, X-ray and radio emissions from the stellar chromosphere, and are thus also associated with the initiation of the mass outflows known generically as stellar winds.

In cool evolved stars the chromosphere represents the interface layer between the photosphere and the wind, and plays a critical role in specifying the amount of mechanical energy that is dissipated as atmospheric heating and deposited as the momentum which results in the initiation of the stellar wind. The chromosphere is key to the mass and energy flux from the entire atmosphere; it also determines the dynamics and magnetic topology of the overlying layers, which contain the wind, and for its part the wind plays a fundamental role in the evolution of stars, especially at the latest phases of their evolution. Therefore, in order to develop realistic theoretical models of stellar winds from cool evolved stars, we need to understand the nature of the dominant physical processes, including the mass and energy flow into the transition region that is positioned between the chromosphere and the wind. The problem of wind formation can therefore be viewed as a problem of chromospheric heating, and thence to its acknowledgement as a problem of fundamental importance for solar and stellar astrophysics.

Chromospheres of cool evolved stars on the red-giant branch (RGB), which are the principal targets of this chapter, represent an extreme case of chromospheres that are observed in cool main-sequence stars. UV observations suggest two distinct classes of cool giants and supergiants, separated in the H–R diagram by the Linsky-Haisch dividing line (Linsky and Haisch 1979). On the blueward (warmer) side of the dividing line, evolved stars earlier than K2 III (also known as 'coronal' giants) show highly compact chromospheres, transition regions and energetic coronae (with

L_X/L_{bol} up to 10^{-3}) that extend into winds that are relatively fast (>100 km s^{-1}), tenuous (<10^{-12} $M_\odot$ year^{-1}), and hot (>1 MK). Those stars are magnetically-active giants; they show variable chromospheric emission and energetic flares which are observed in the UV, EUV and X-ray regions. Direct observations of surface magnetic fields in over 60 active giants have produced measurements of field strengths that can be as much as 100 G (Konstantinova-Antova et al. 2012). The redward (cooler) side of the dividing line is characterized by late-K and M-type giants (the 'non-coronal' giants); they show relatively weak transition-region signatures but do have extended chromospheres (0.16–0.4 stellar radii) and massive slow winds (from 10^{-11} to 10^{-6} $M_\odot$ year^{-1}), whose typical velocities of $\leq$40 km s^{-1} are considerably smaller than the expected escape velocities at the stellar surface. Intermediate between the coronal and the non-coronal giants are the so-called 'hybrid' giants, which show signatures of both compact chromospheres and pronounced transition regions; those transition regions are characterized by emission from N V, C IV and Si IV, hot coronal plasma (1–20 MK) observed in X-rays, and winds that are hot ($T \sim 1$ MK), tenuous and fast (>100 km s^{-1}) with associated mass-loss rates of 10^{-11} $M_\odot$ year^{-1}.

Despite decades of observational and theoretical studies, the physical nature of atmospheric heating and associated mass loss is not well understood. (a) *What physical mechanism accounts for the existence of three types of cool evolved stars?* (b) *What processes control the extent of the chromospheres in those stars, and their mass-loss rates?* (c) *What physical processes are involved in heating a stellar atmosphere and accelerating the star's wind?* To address these questions, any theoretical model of outer atmospheres of cool stars should describe an intimate relationship between the heating of the chromosphere, the transition region and the corona, and the heating and acceleration of the wind.

From that perspective, K5–M giants and supergiants, and in particular those that are in eclipsing binaries, serve as an ideal laboratory for investigating how a relatively simple atmosphere that consists of an extended chromosphere of $0.2-1$ stellar radii ($R_\star$) develops into a slow and massive stellar wind. In this chapter we highlight recent progress in our understanding of chromospheres and winds from studies of cool evolved stars, building on previous reviews of chromospheric heating processes by (inter alia) Narain and Ulmschneider (1990, 1996). In relation to the atmospheres and winds of cool evolved stars, we discuss theoretical constraints to heating and acceleration that are derived from semi-empirical models, and go on to describe the successes that have been achieved by accounting for atmospheric heating via acoustic waves, but mention also the attendant limitations. We also highlight recent approaches to modelling wind acceleration from cool stars from the aspect of magnetohydrodynamic (MHD) waves, and conclude by discussing possible future developments in theoretical modelling.

5.2 Observational Constraints to the Heating and Acceleration of Stellar Atmospheres and Winds

As in the case of the chromosphere of the Sun, the chromosphere of a cool evolved star consists of highly complex, time-dependent, optically thick, weakly ionized, magnetized atmospheric layers. Dynamic chromospheres of non-coronal giants and supergiant stars transition from quasi-static atmospheres (in the most simplistic case) into stellar winds. Models of such complex multi-parameter environments need to be based on observational constraints. Various sophisticated multi-dimensional models that have recently been developed attempt to describe the chromosphere of the Sun, and—by implication—stellar chromospheres and winds as well (Suzuki 2007; Hansteen et al. 2010; Airapetian et al. 2014). To make those models more realistic, many input parameters are needed to describe the sources and the specifics of the mechanical energy flux that is generated within the solar or stellar photosphere. They also require reference multi-dimensional model atmospheres. There is a variety of methods by which realistic theoretical models can be constrained.

Extensive observations of samples of G–M giants and supergiants—for instance, by Carpenter et al. (1994, 1995), Brown et al. (1996), Reimers et al. (1996), Robinson et al. (1998), Ayres et al. (1998, 2003), Ayres (2005), Dupree et al. (2005), Harper et al. (2005, 2013), Harper (2010), and Pérez Martínez et al. (2011)—provide important clues to the thermodynamics and kinematics of the chromospheres and winds of the various stellar types. Recent high-spectral-resolution aperture-synthesis imaging of two supergiants, Antares and Betelgeuse, has revealed an asymmetry and inhomogeneity of their chromospheric structures and the existence of a clumpy, cool, outer molecular shell, the 'MOLsphere' (Tsuji 2000), extending out to 1.2–1.5 stellar radii (Ohnaka 2013; Ohnaka et al. 2013).

In considering the atmospheric plasmas of stars, we can divide the observational constraints into two major categories: energy dissipation requirements and momentum deposition requirements.

5.2.1 *Energy Dissipation Requirements*

In the absence of significant flows, the dissipation of chromospheric energy due to non-radiative energy source(s) is mostly balanced by radiative cooling. The observed surface fluxes of the two major contributors, i.e., the Mg II and Ca II emission lines, allow one to define the range of required heating rates. Those have been given as $(1–100) \times 10^5\,\mathrm{ergs\,cm^{-2}\,s^{-1}}$ (Linsky and Ayres 1978; Strassmeier et al. 1994; Pérez Martínez et al. 2011).

One-dimensional semi-empirical models of evolved stars represent powerful tools for constraining the radial profiles of the heating rates that are related to the deposition of energy throughout the atmosphere. This class of model was inspired by time-independent 1-D semi-empirical models of the solar chromosphere developed by Vernazza et al. (1976, 1981) and Fontenla et al. (1990, 2002); they were designed to reproduce the temporally and spatially averaged UV line profiles and fluxes. Semi-empirical models provide a quantitative characterization of the radial profiles of temperature, electron density, neutral hydrogen density and turbulent velocity across the atmospheres of evolved stars. This type of model was developed for a number of evolved stars, such as giants like α Boo, α Tau, and β Cet, and for various supergiants, including the eclipsing supergiant 31 Cyg (Ayres and Linsky 1975; Eriksson et al. 1983; McMurry 1999; Eaton 2008). A chromospheric model for α Tau developed by McMurry (1999) suggests that the temperature rises throughout the chromosphere up to 100,000 K at about 0.2 $R_\star$. At the same time, the chromosphere transitions into a wind within one stellar radius, suggesting that the atmosphere therefore undergoes acceleration between 0.2 and 1 $R_\star$. *FUSE* observations of various non-coronal giants show the presence of C III and O VI lines, indicating hot plasma with temperatures up to 300,000 K. Those very observations in 22 Vul were described in Sect. 2.4.5.2 and Fig. 2.24. Plasma at such high temperatures occupies low volumes and appears to be mostly at rest with respect to the photosphere in stars that have winds of low escape velocities, indicating that the plasma should be magnetically confined (Ayres et al. 2003; Harper et al. 2005; Carpenter and Airapetian 2009).

Recent detections of surface magnetic fields for some G–M giants and supergiants suggest that they could be an important contributor to the thermodynamics of the outer chromosphere (Auriére et al. 2010; Konstantinova-Antova et al. 2010, 2012). The observed field strengths vary from 0.5 to 1.5 G in late-type giants and increase to 100 G in early-type coronal giants. Rosner et al. (1995) suggested that, as stars evolve towards the giant phase, their magnetic topology transitions from closed magnetic configurations to predominantly open ones; the latter allow massive, non-coronal winds to be supported. If the magnetic field is non-uniformly distributed over the stellar surface, the associated radial profiles in the atmosphere can be determined by assuming that the magnetic pressure, $B^2/8\pi$, inside an untwisted (purely longitudinal) flux tube is balanced by the gas pressure of the surrounding non-magnetic atmosphere, P_{ext}. This suggests that the plasma pressure inside the tube is smaller than the magnetic pressure of the plasma, $\beta = 7n_9T_4/B_1^2$, where $n_9 = n/10^9\,\text{cm}^{-3}$, $T_4 = T/10{,}000\,\text{K}$, and $B_1 = B/10\,\text{G}$. For typical chromospheric conditions of $n_9 \sim 1$ and $T_4 \sim 1$, the plasma-β becomes less than 1 at $\text{B} \geq 50\,\text{G}$. Observations in the vicinity of active regions on the Sun that are represented by plages indicate magnetic fields of a few hundred Gauss at chromospheric densities and temperatures; the force balance between the magnetic and plasma pressures can therefore be described satisfactorily by the thin flux-tube approximation (Rabin 1992; Gary 2001; Steiner 2007; Judge et al. 2011). The vertical profile of the chromospheric magnetic field can therefore be determined as

$$B_z(z) = \sqrt{8\,\pi\,P_{\mathrm{gas}}}. \tag{5.1}$$

Once the magnetic field is known, the profile of the Alfvén velocity, V_A, can be calculated throughout the chromosphere as

$$V_A = \frac{B_z(z)}{\sqrt{4\,\pi\,\rho(z)}}, \tag{5.2}$$

where $\rho(z)$ is the mass density.

Since the photospheres of giants and supergiants are convective and dense, photospheric footpoints of longitudinal magnetic fields are forced to follow the convective motions within the photosphere. The motions of magnetic field lines with a frequency of the inverse turnover time of a stellar granule, $\nu_A = H_{\mathrm{P}}/V_c$, with H_{P} as photospheric pressure scale height and V_c as convective velocity, are able to excite MHD waves along or across the magnetic flux tube, including torsional or transverse Alfvén waves (Ruderman et al. 1997). Torsional Alfvén waves (Alfvén 1942) represent linearly incompressible azimuthal perturbations of the plasma velocity (linked to the azimuthal perturbations of the magnetic field) that, unlike compressible waves such as longitudinal MHD waves, do not disturb the plasma density. Although Alfvén waves were predicted in 1942, it is only relatively recently that researchers have reported the observational detection of them in the solar chromosphere and corona (Tomczyk et al. 2007; De Pontieu et al. 2007; Jess et al. 2009).

Alfvén waves launched from the stellar photosphere propagate upwards into a gravitationally stratified atmosphere and are subject to reflection from regions of high gradients of Alfvén velocity if the wave frequency, ν_A, is less than the critical frequency, $\nu_{\mathrm{crit}} = dV_A/dz$ (Heinemann and Olbert 1980; An et al. 1990). The interaction of downward-reflected Alfvén waves with upward propagated ones can ignite a turbulent cascade of Alfvén waves in the lower solar atmosphere and provide a valid source for heating the solar corona and stellar winds in open field regions (Matthaeus et al. 1999; Cranmer and Ballegooigen 2005; Cranmer and Saar 2011).

Reflection of Alfvén waves can play an important role in driving slow and massive winds from giants and supergiants (An et al. 1990; Airapetian et al. 1998, 2000, 2010; Suzuki 2007; Cranmer and Saar 2011). The radial profile of the critical frequency therefore provides important information about the role of the heating and momentum deposition of Alfvén waves in the atmosphere. The critical Alfvén frequency can be calculated directly from a semi-empirical model by differentiating the Alfvén velocity profile given by Eq. 5.2. By applying that procedure to the case of a K5 giant (α Tau) using the semi-empirical model of McMurry (1999), one can show that the Alfvén velocity gradient reaches its maximum at $0.21\,R_\star$. Figure 5.1 suggests that waves at frequencies less than 0.6 mHz are trapped in the chromosphere of α Tau (left panel). The same technique, based on the semi-empirical model developed by Eaton (2008) and discussed in Sect. 2.4.2.2, can also be applied to calculate the gradient of the Alfvén velocity for the K5 supergiant

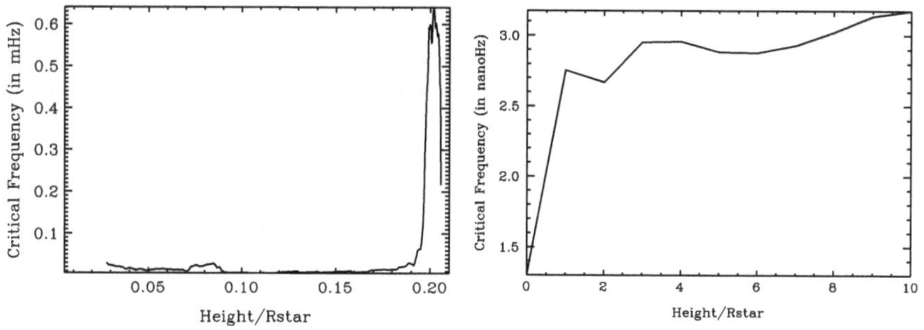

Fig. 5.1 Radial profiles of the critical frequency in a solar or stellar chromosphere as predicted by semi-empirical models of α Tau (*left*) and 31 Cyg (*right*)

primary in 31 Cyg. The right panel of Fig. 5.1 shows the vertical profile of dV_A/dr in the chromosphere of 31 Cyg, and suggests that waves at $\nu \leq 3$ nHz should be trapped in the first 10 stellar radii.

The magnetic field and the Alfvén velocity profile can also be probed by using the Poynting theorem (Jackson 1999):

$$\frac{\partial W}{\partial t} + \nabla \cdot \mathbf{S} = -\mathbf{E} \cdot \mathbf{J}, \tag{5.3}$$

where $W = \frac{1}{8\pi}(E^2 + B^2)$ is the electromagnetic energy density and $\mathbf{S} = \frac{1}{4\pi}\,\mathbf{E} \times \mathbf{B}$ is the Poynting vector of the energy source. $\mathbf{S}$ represents the Poynting flux of Alfvén waves launched from the photosphere. For a steady-state chromosphere, $\frac{\partial W}{\partial t} = 0$ and $\mathbf{S}$ has only an upward component S_z. We thus obtain

$$\frac{dS_z}{dz} = - < q >, \tag{5.4}$$

where $< q >$ is the time-averaged heating rate at a given height, z (see also Song and Vasyliunas 2011).

The heating rate of the plasma can be derived from the energy equation for a steady-state chromosphere in which the heating rate is balanced by the thermal conductive and radiative cooling rates, referred to as $L_{\rm cond}$ and $L_{\rm rad}$, respectively. It is found that

$$< q >= L_{\rm cond} + L_{\rm rad}. \tag{5.5}$$

In a stellar chromosphere, $T < 0.5$ MK; the thermal conduction time is much longer than the radiative cooling time, and the thermal conduction cooling term can be safely neglected. Consequently, the radial profile of the observationally derived cooling rates provides direct clues about the profile of the Poynting flux of the heating energy source. Detailed information about the radial profiles of the chromospheric magnetic field and the Alfvén velocity can be obtained if it is assumed that Alfvén waves are the major source of the chromospheric heating. Observations of chromospheric lines that are broadened, but not by thermal processes, also imply that Alfvén waves may be the dominant source of energy and wind acceleration in cool giants and supergiants; further relevant discussions have been given by Airapetian et al. (2010) and Cranmer and Saar (2011). This type of incompressible transverse wave can be excited directly, presumably through the shuffling or twisting of magnetic flux tubes by well-developed magneto-convection in stellar photospheres (Ruderman et al. 1997; Musielak and Ulmschneider 2002). Recently, Morton et al. (2013) presented observational evidence that incompressible waves can be excited in the solar photosphere by the vortex motions of a concentration of magnetic flux.

The energy flux of Alfvén waves excited at the photosphere is defined by the z-component of the Poynting vector. By applying Ohm's law, $\mathbf{E} = \eta\mathbf{J} - \mathbf{V} \times \mathbf{B}$, Ampere's law, $\mathbf{J} = \frac{1}{4\pi}\nabla \times \mathbf{B}$, and using vector identities, we can write the upward Poynting flux in Alfvén waves as

$$\mathbf{S} = \frac{1}{4\,\pi}[\mathbf{V}\,B^2 - \mathbf{B}(\mathbf{V}\cdot\,\mathbf{B})] + \frac{\eta}{4\pi}(\nabla \times \mathbf{B}) \times \mathbf{B}. \tag{5.6}$$

If we further assume the existence of the azimuthal component only of the velocity of footpoint motions, $V_\phi \neq 0$, i.e., that there are no vertical motions in the photosphere (so $V_z = 0$), and if we represent the total magnetic field as the sum of the background longitudinal flux-tube magnetic field B_z plus the perturbed field δB due to Alfvén waves, we obtain the z-component of the upward Poynting flux as

$$S_z = -\frac{1}{4\,\pi}\,B_z\,V_\phi\,\delta B - \frac{\eta}{4\,\pi}\delta B\frac{\partial\delta B}{\partial z}. \tag{5.7}$$

For high magnetic Reynolds numbers, $\mathrm{Re}_m = \frac{V_A L}{\eta}$ (η is the magnetic diffusivity) and is >10 in a stellar chromosphere, so the second term in Eq. 5.7 can be neglected with respect to the first one. Then, following the Wallén relation $\delta V_A = \delta\frac{B}{\sqrt{4\pi\rho}}$ and assuming that waves are incompressible (so $\delta\rho = 0$), we obtain

$$\frac{\delta V}{V_A} = \frac{\delta B}{B_z}. \tag{5.8}$$

This assumption is valid until Alfvén waves become strongly non-linear and convert a significant fraction of their energy into longitudinal waves (Ofman and Davila 1997; Suzuki 2007; Airapetian et al. 2014). By substituting δB from Eq. 5.8 into Eq. 5.7, we obtain the Poynting flux as

$$S_z = \rho\,<\delta V^2>\,V_A. \tag{5.9}$$

Furthermore, when combining Eqs. 5.4, 5.5 and 5.9, we obtain the following:

$$\frac{d}{dz}(\rho < \delta V^2 > V_A) = -L_{\mathrm{rad}}(z), \tag{5.10}$$

Equation 5.10 relates the thermodynamic quantities such as the plasma density, turbulent velocity and the radiative cooling rates, which are obtained from semi-empirical models, to the hitherto unknown vertical profile of the Alfvén velocity. Equation 5.10 can be rewritten as:

$$e_A \frac{dV_A}{dz} + \frac{de_A}{dz} V_A = -L_{\mathrm{rad}}, \tag{5.11}$$

where $e_A = \rho < \delta V^2 >$ is the energy density of Alfvén wave energy.

Once V_A is known, the profile of the magnetic field throughout the chromosphere can be determined. *Hence, the knowledge of V_A and subsequent retrieval of $B_z(z)$ represents the missing link between thermodynamic-based semi-empirical models and MHD-based theoretical models of chromospheres and winds.* This last equation allows us to determine the range of critical frequencies at which Alfvén waves become reflected from regions where the Alfvén velocity gradient is at a maximum.

Comparing the magnetic-field profiles derived from Eq. 5.11 with the one obtained from Eq. 5.1 enables us to determine the degree of deviation of the magnetic field in a chromosphere from the longitudinal (untwisted) magnetic field, thus allowing us to constrain the value of the azimuthal magnetic field. The magnetic-field profile in the chromosphere of α Tau as derived decreases with height at the rate of a super-radial expansion factor, $f(r)$. Then, the magnetic field varies with the height, r, as $B(r) \sim f(r)/r^2$, which is less steep than the profile obtained by Kopp and Holzer (1976) for solar coronal holes.

The next generation of semi-empirical models of evolved stars should therefore combine high-resolution spectroscopic and spatial information. Eclipsing binaries offer a unique opportunity to derive geometric constraints to the observed chromospheres and their winds (see Eaton et al. 2008, and Chap. 2, particularly Sect. 2.8). Another promising approach utilizes high spatial-resolution interferometric observations of various giant and supergiant stars.

5.2.2 *Momentum Deposition Requirements*

The momentum deposition from non-radiative energy sources into stellar winds should explain the observed energy fluxes. An estimate of the energy flux that should be added in order to drive a steady-state wind above the atmospheric base can be derived from the energy equation for a steady-state wind in a single-fluid hydrodynamic approximation:

$$F_{\text{wind}} = \frac{(-\dot{M})}{4\pi\, R_\star^2}\,(V_\infty^2 + V_{\text{esc}}^2). \tag{5.12}$$

V_∞ is the wind terminal velocity, and V_{esc} is the escape velocity of the star at its surface (Holzer 1987). Detailed examination of chromospheric emission lines of Fe II, O I and Mg II indicate that the wind from a giant appears to originate near the base of the chromosphere and it continues to accelerate throughout the entire chromospheric region (Carpenter et al. 1995). It is therefore assumed that the wind reaches its terminal velocity within one stellar radius.

Table 5.1 presents radii, $R_\star/R_\odot$, escape velocities, V_{esc}, wind velocities, V_∞, and mass loss rates, $\dot{M}$, for selected stars (Robinson et al. 1998; Eaton 2008; Harper 2010; Neilson et al. 2011). A significant portion (from 65 to 95%) of the mechanical energy is required to lift the plasma beyond the gravitational well of the star and form the wind.

The energy fluxes required to generate the winds from α Tau, α Ori and 31 Cyg are 2.8×10^3, 2.2×10^5 and 2.5×10^5 ergs cm^{-2} s^{-1}. This suggests that, in non-coronal giants, F_{wind} is only a few percent of the outer atmospheric heating rate, and therefore over 90% of the energy flux of the mechanical source is trapped in the chromosphere while only a small fraction leaks out to accelerate and heat the wind. For supergiants the wind is initiated in much higher regions of the extended atmosphere; it reaches its terminal velocity before leaving the chromosphere (see Carpenter and Robinson 1997).

Another requirement for models of stellar winds arises from the constraints to the terminal wind velocities and mass-loss rates. For example, non-coronal giants and supergiants, including those in ζ Aur systems, show evidence of the presence of winds that are massive and slow—a few tens of km s^{-1}. Models of steady-state, spherically-symmetric winds suggest that, in order to produce the low terminal velocities (ones that are less than half the surface escape velocities) and the high mass-loss rates, most of the energy and momentum should be deposited below the sonic point, while momentum addition beyond the sonic point should produce the fast and tenuous solar-like winds (Hartmann and MacGregor 1980; Holzer 1987).

Table 5.1 Wind properties of α Tau (K5 III), α Ori (M2 Iab) and 31 Cyg (K4 Ib)

Star	$R_\star/R_\odot$	V_{esc} (km s^{-1})	V_∞ (km s^{-1})	$\dot{M}$ ($M_\odot$ year^{-1})
α Tau	44	115	30	1.6×10^{-11}
α Ori	955	64	10	2×10^{-6}
31 Cyg	197	152	90	3×10^{-8}

5.2.3 Constraints from Atmospheric Turbulence and Flows

High-resolution spectroscopy of many cool giants and supergiants has demonstrated that the profiles of optically thin UV lines of C II], Si II, Si IV and C IV that arise from the chromosphere and transition regions in those stars show direct evidence of strong turbulence. The C II] line is formed at temperatures between 5000 and 10,000 K, and is optically thin in all of the stars observed. It has an intrinsically narrow profile, so the line width primarily reflects the Doppler broadening; the enhanced wings of the line profiles can be simulated by a single Gaussian profile (Carpenter et al. 1991). The deduced turbulent velocities range from 24 km s^{-1} for the K5 giant α Tau to 35 km s^{-1} for the M1.5 supergiant α Ori, thus suggesting supersonic turbulence in their chromospheres; further discussions have been given by Cuntz (1997), Robinson et al. (1998), and Cuntz et al. (2001). Many of the UV line profiles can be fitted by a combination of narrow and broad Gaussian components, as discussed by Wood et al. (1997), Dupree et al. (2005) and Eaton (2008). For example, fitting the profiles of C II] in the M3.5 III star γ Cru requires a two-component Gaussian model with FWHM $= 27$ and 42 km s^{-1}, while for α Ori those values need to be 19 and 48 km s^{-1} (Eaton 2008).

Hybrid and coronal giants show much larger non-thermal velocities in transition region lines, reaching 200 km s^{-1}, and also show both narrow and broad Gaussian components. Moreover, the non-thermal broadening of UV lines observed in quiescent spectra of coronal giants and in active dwarfs, including the Sun, tends to increase with temperature (Ayres et al. 1998; Pagano et al. 2004; Peter 2006).

The Sun also shows non-thermally broadened, two-component Gaussian-shaped, red-shifted UV lines that form in the transition region (Peter 2006). That study does not confirm the idea suggested by Wood et al. (1997) and Pagano et al. (2004) that the broad component is heated by microflares. In contrast, the spectrum of the Sun's integrated disk was used to demonstrate that the broad components are a consequence of the structure of its chromosphere. Carpenter and Robinson (1997) suggested that broad components may be a misleading description of the physics of a stellar chromosphere; instead of fitting the profiles by two Gaussian curves, they explain the enhanced wings by a signature of large-scale turbulence, which is anisotropically distributed along the line of sight and is directed preferentially either along, or perpendicular to, the radial direction. Airapetian et al. (1998, 2000, 2010) developed a 2.5-D MHD model of a stellar wind in which supersonic turbulent motions that are responsible for non-thermal broadening in the UV lines can be attributed to unresolved motions of upward-propagating non-linear Alfvén waves. The anisotropy formed by a large-scale longitudinal (open) magnetic field in an atmosphere can contribute to the formation of broad wings such as those observed in evolved giants and supergiants. Another important feature of the outer atmospheres of cool stars like the primary of 31 Cyg is the clumping of the gas, as proposed by Eaton (2008) and discussed further by Harper (2010).

Optically thin chromospheric lines from these stars and the Sun also exhibit a nett red-shift. This red-shift is indicative of downward motions of a few $\mathrm{km\,s^{-1}}$ in non-coronal giants, and up to $30\,\mathrm{km\,s^{-1}}$ in coronal giants as observed in lines from the chromosphere and transition region (Ayres et al. 1998; Robinson et al. 1998; Peter and Judge 1999; Doyle et al. 2002; Eaton 2008). It is interesting to point out that Doyle et al. (2002) have shown observationally that the red Doppler shifts in the solar emission lines of O III, O IV and Ne VIII exhibit short variability of a few minutes. That timescale is comparable to the characteristic period of a few minutes for MHD waves observed in the solar chromosphere (Morton et al. 2012). Assuming that MHD waves are excited by magneto-convective motions with the turbulent turnover time of a convective cell, $\tau = L_g/V_{\mathrm{conv}}$, we can expect a variation of red-shift shifts in chromospheric lines of red giants with a timescale of a few days. Judge and Carpenter (1998) concluded that such turbulent motions indicate downward-propagating non-linear waves in the chromosphere. 3-D MHD models of the transition region in the Sun also interpret the observed emission-line red-shifts in terms of downward-propagating compressive waves (Hansteen 1993), or of material heated within low-lying transition-region loops that later cool and fall down into the chromosphere (Guerreiro et al. 2013). This process can be very important in the chromospheric dynamics and energy balance, and therefore provides a clue to the solar and stellar atmospheric heating. Can those models be incorporated into describing the mechanisms for heating chromospheres and winds? The non-thermal broadening and red-shifts of chromospheric lines thus represent one of the major signatures of chromospheric heating that need to be addressed by any valid theoretical model.

5.3 Acoustic Heating: Successes and Limitations

5.3.1 Two-Component Chromosphere Models

Following previous reviews, for instance by Narain and Ulmschneider (1990, 1996) and references therein, particularly the work by Schrijver (1987) and Rutten et al. (1991), it has become obvious that, from a general point of view, stellar chromospheres can be considered as consisting of acoustically-heated *and* magnetically-heated components. The general heating rate, proliferated to stellar chromospheres including those of non-coronal giants and supergiants, can therefore be expressed as

$$F = F_{\mathrm{ac}}(T_{\mathrm{eff}}, g_\star, Z) + F_{\mathrm{mag}}(T_{\mathrm{eff}}, g_\star, Z, P_{\mathrm{rot}}), \qquad (5.13)$$

where F_{ac} is the acoustic heating rate and F_{mag} is the magnetic heating rate. Both increase as a function of stellar effective temperature T_{eff}, increase with decreasing stellar surface gravity $g_\star$ (i.e., by a few orders of magnitude between main-sequence stars and low-gravity supergiants), and show some dependence on the stellar metallicity Z, though it is much less important than the impact of $g_\star$.

The magnetic heating rate also depends—at least in a statistical sense—on the rotation rate of the star, such that it is lower for slow rotating (i.e., older) stars, especially giants and supergiants, which have evolved away from the main-sequence (Schrijver 1993). That dependence has profound consequences for the resulting amounts of chromospheric emission (e.g., Skumanich 1972; Noyes et al. 1984; Simon et al. 1985; Strassmeier et al. 1994), as identified in multiple spectral regimes, including detailed observations of Ca II and Mg II. The latter have been interpreted through empirical, semi-empirical and theoretical concepts, including statistical relationships, and have also been utilized to decipher information about the thermodynamic, (magneto-)hydrodynamic and radiative properties of stars at different ages and evolutionary status. Detailed analyses focused on non-coronal giants have been described by Schrijver (1993), Schrijver and Pols (1993), and others. Theoretical studies have been carried out by MacGregor and Charbonneau (1994) and Charbonneau and MacGregor (1995) into the link between stellar atmospheric and wind properties on the one hand, and the evolution of stellar dynamos on the other; this link constitutes the physical reason for the fundamental changes like the transition from coronal main-sequence stars to non-coronal giants.

Another pivotal aspect concerns the study of one-component (acoustic only) and two-component (acoustic plus magnetic) simulations of chromospheric heating itself. Detailed theoretical models for stars of different spectral types, and for non-coronal giants, have been calculated by Ulmschneider (1989), Buchholz et al. (1998), Cuntz et al. (1998, 1999), and Fawzy et al. (2002). Buchholz et al. (1998) presented 1-D time-dependent models of acoustically-heated chromospheres for main-sequence stars between spectral types F0 V and M0 V, and for two giants akin to spectral types K0 III and K5 III. The emergent radiation in Mg II *h* and *k* and Ca II *H* and *K* was calculated and compared with observation. They found good agreement across nearly two orders of magnitude between the time-averaged emission in those lines and the observed basal flux emission, which had been suspected as due to non-magnetic (i.e., acoustic) heating operating in all late-type stars. The authors pointed out that their results clearly supported the idea that the 'basal heating' of the chromospheres of late-type stars and of non-coronal giants (see Fig. 5.2) is due to acoustic waves. The latter result has been disputed in the meantime (Judge and Carpenter 1998) from the argument that acoustic-wave heating, in the context of existing models, is unable to explain the levels of turbulence that are observed (see Sect. 5.2.3).

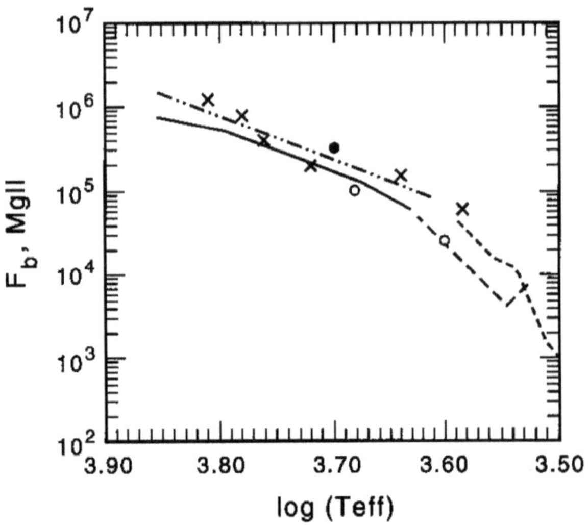

Fig. 5.2 The basal flux limit for Mg II *h* and *k*, shown as a function of T_{eff}, is compared to results derived from observations. The *solid line* represents the results of Rutten et al. (1991), and the *double-dotted/dashed line* the earlier ones from Schrijver (1987). The *long-dashed line* shows the upper limits derived for very cool dwarfs (Doyle et al. 1994), while the *short-dashed line* is the Mg II basal flux limit for giants (Judge and Stencel 1991). The *crosses* and *open circles* represent theoretical results from Buchholz et al. (1998), for main-sequence stars and for giants, respectively. The *closed circle* at $\log T_{\mathrm{eff}} = 3.70$ is the theoretical result of Cuntz et al. (1994), but ignores the small differences in the Mg II flux due to the different atmospheric metallicities considered *(Reproduced, with kind permission, from Buchholz et al. (1998))*

5.3.2 Possible Relevance of Acoustic Waves to Winds from Cool Evolved Stars

An important feature of acoustic waves in stellar atmospheric environments, including the chromospheres of evolved giants and supergiant stars, is that they dissipate the lion's share of their mechanical energy flux fairly close to the stellar photospheres; Ulmschneider (1989) compared the dissipative behaviour of acoustic waves in giants and dwarfs of identical effective temperatures. That key result concerning the dissipative properties of acoustic waves in cool evolved stars, implying that acoustic waves cannot support the massive winds of those stars, has also been discussed in the broader context of proposed stellar mass-loss mechanisms by Holzer (1987). It entails an additional justification for the exploration of Alfvén wave-driven winds for non-coronal evolved stars as opposed to non-magnetic mechanisms. Models of Alfvén wave-driven winds were pursued by Hartmann and MacGregor (1980), Hartmann and Avrett (1984), and more recently by Airapetian et al. (2014). A detailed study by Cuntz (1990), focused on α Boo (K2 III) and based on adequate acoustic-wave frequency spectra, yielded acoustically-initiated mass-loss rates of the order of 10^{-13}–10^{-15} $M_\odot$ year^{-1}, which are more than a factor of 10^4 *below* the limit established observationally.

A decisive shortcoming of acoustic waves is that they fail to dissipate their mechanical energy over a significant distance that is comparable to the height of the critical point (Holzer and MacGregor 1985; Cuntz 1990; Judge and Stencel 1991). That property then leads to unacceptably low mass-loss rates. Previous relatively positive assessments about the ability of acoustic waves to drive mass loss from evolved stars were given by Pijpers and Hearn (1989). They explored a simple stellar wind model, loosely guided by the stellar parameters of α Ori, which led to mass-loss rates in the range 10^{-8}–10^{-4} $M_\odot$ year^{-1}, depending on the wave parameters selected, and at first sight it appeared to be promising. However, that type of model must be regarded as inapplicable because it is based on the assumption that the acoustic waves are time-independent and linear and thus exclude the possibility for shock formation, for example. The results that were obtained in that way then are not meaningful.

Nevertheless, acoustic waves are still expected to be relevant for initiating mass loss from non-coronal giants and supergiant stars since they are able to increase the thermal pressure and density scale heights in a spatially-extended chromosphere owing to their dissipative behaviour such as heating and transfer of wave pressure; see Buchholz et al. (1998) and references therein. Another possible contribution of non-magnetic processes, pointed out by Suzuki (2007) for example, is the initiation of turbulent pressure that may help to lift matter closer to the critical point of the stellar wind. Nevertheless, detailed assessments, in the framework of 1-D models for acoustically-generated turbulence in chromospheric models of α Tau, show that the synthetic results are significantly lower than those measured in GHRS observations (Judge and Cuntz 1993). The question as to whether that discrepancy is due to the unrealistic 1-D assumption of existing theoretical models, or whether it is more profoundly enshrined in the basic physics of the acoustic heating mechanism, is still an unsettled debate (see Judge and Carpenter 1998; Cuntz et al. 2001). Preliminary 3-D models of convectively-initiated turbulence in α Ori have been given by Freytag et al. (2002).

5.4 MHD Wave-Heating and Wind Acceleration

5.4.1 Energy Dissipation Due to Alfvén Waves: A Source of Chromospheric Heating

Magnetic heating mechanisms for solar and stellar chromospheres have been targeted in numerous reviews, including those by Narain and Ulmschneider (1990, 1996). Two major types of heating mechanisms have generally been proposed, which are broadly classified as AC (alternating current, such as MHD wave dissipation) and DC (direct current, such as magnetic-field dissipation through magnetic reconnection). In this section we focus on the progress regarding AC heating processes and their observational signatures. MHD-wave heating can be driven by two types of waves: compressible longitudinal MHD waves (slow and fast magneto-sonic waves), and incompressible transverse waves (torsional Alfvén waves). According to Ulmschneider et al. (2001) and Musielak and Ulmschneider (2002), the energy fluxes of longitudinal and transverse waves in cool evolved stars are comparable, of the order of 10^7–10^8 ergs cm^{-2} s^{-1}. That amount of energy generated by waves in the photospheres of cool giants is sufficient to account for the observed cooling rates together with the energy needed to drive winds (see Sect. 5.2.1).

Torsional Alfvén waves have been suggested as a likely source of the heating of the solar chromosphere and corona (Osterbrock 1961; Hollweg 1973, 1978; Heinemann and Olbert 1980; Holzer et al. 1983; Cranmer and Ballegooigen 2005; Mathioudakis et al. 2013). The same approach has been extended to the outer atmospheres of cool evolved stars by Hartmann and MacGregor (1980), Holzer et al. (1983), Hartmann and Avrett (1984), Suzuki (2007, 2013), and Cranmer (2008, 2009).

Alfvén waves can deposit energy into solar and stellar atmospheres through a number of mechanisms. For example, in closed magnetic structures resonant absorption mechanisms may become efficient, while in closed and open structures energy dissipation through the cascade due to Alfvén-wave turbulence or mode conversion may become efficient (Davila 1987; Matthaeus et al. 1999; Cranmer and Ballegooigen 2005; Suzuki and Inutsuka 2005; Mathioudakis et al. 2013; Airapetian et al. 2014). Studies by An et al. (1990), Barkhudarov (1991), Rosner et al. (1991), Velli (1993), MacGregor and Charbonneau (1994) and Charbonneau and MacGregor (1995) concluded that, as waves propagate in a gravitationally-stratified atmosphere, they may become subject to reflection from atmospheric regions where the gradient in the Alfvén velocity is comparable to, or greater than, the Alfvén wave frequency.

While those studies point to the possible importance of magnetic-wave pressure in chromospheric heating, they suffer from the restrictive nature of assumptions of linearity and from the fact that they are not solving consistently the relevant MHD equation involving the magnetic field, density and velocity. Although the linear treatment of winds in cool, luminous stars has shown that MHD turbulence can be important for driving the winds, those models are incapable of examining properly the wave dissipation, which is a crucial part of the mechanism. Furthermore, because the entire set of non-linear time-dependent MHD equations is not solved consistently in those studies, there is the possibility that important physical effects have been neglected or overlooked. An important example is that the coupling between the azimuthal and radial components of the velocity and magnetic fields are only treated in a linear approximation. Recent models by Suzuki (2007, 2013) treat self-consistently the dissipation of Alfvén waves in forming stellar chromospheres and coronal layers that expand into winds, but they assume that Alfvén waves are launched from a fully-ionized photosphere—an approximation that is not applicable, since the degree of ionization is by no means uniform; the atmosphere of a giant or supergiant is a combined system of photosphere plus chromosphere, and the actual degree of ionization in most cases graduates from trivially small in the photosphere to nearly total at the top of the chromosphere. The treatment of the stellar photosphere and chromosphere as a fully-ionized plasma therefore overlooks a wide range of physical effects of current dissipation in a partially-ionized magnetized plasma (Goodman 2000; Khodachenko et al. 2004).

In a weakly-ionized medium such as a stellar chromosphere, where there is varying collisional coupling throughout between ions and electrons, new effects can appear, such as the non-ideal Hall effect or ambipolar diffusion. In astrophysics, ambipolar diffusion usually refers to the decoupling of neutral and charged particles in a plasma. If both electrons and ions are magnetized, or frozen-in into the magnetic field, neutral particles do not 'feel' the magnetic field and will slip through it. Neutral particles then drag ions with them via ion–neutral collisions, introducing an electric field perpendicular to the magnetic field lines. The Hall effect occurs when electrons are magnetized but ions are not. In such a case the Hall electric field results from the drift velocity of electrons with respect to ions, because the two kinds of charged particles respond differently to collisions from the neutral particles. In solar and stellar atmospheres the ambipolar diffusion perpendicular to the magnetic field is many orders of magnitude greater than the classical Spitzer diffusion along the magnetic field. The collisional coupling between charged particles and neutral gas is therefore a fundamental process in weakly-ionized and strongly-magnetized solar and stellar chromospheres and winds (Piddington 1956; Osterbrock 1961; Hartmann and MacGregor 1980; Holzer et al. 1983). Many recent studies (Goodman 2000, 2004; De Pontieu et al. 2001; Khodachenko et al. 2004; Leake et al. 2005; Krasnoselskikh et al. 2010; Soler et al. 2013; Tu and Song 2013) have shown that in the solar chromosphere, which is a weakly-ionized and magnetized atmosphere, the effect of ion–neutral collisions becomes significant in dissipating the electric currents that are introduced by MHD waves.

Airapetian et al. (2014) found that the effects of ambipolar diffusion may also play a critical role in the chromospheres of cool giants. It is known that the photospheres of cool giants and supergiants are characterized by well-developed magneto-convection, with representative velocities up to $10\,\mathrm{km\,s^{-1}}$ (Gray 2008; Chiavassa et al. 2010). Interaction of such motions with open magnetic fields may excite longitudinal and transverse MHD waves. What happens is that transverse Alfvén waves cause periodic fluctuations of plasma motions perpendicular to the magnetic field lines and thereby induce an electric field, $\mathbf{E} = \delta\mathbf{V} \times \mathbf{B}/c$, in the reference frame of the plasma. This induced electric field generates electric currents. The dissipation of the currents induced by the Alfvén waves perpendicular to the magnetic field provides an efficient source for converting the kinetic energy of convection into electrical energy. Working from the approach of Braginskii (1965), De Pontieu and Haerendel (1998), De Pontieu et al. (2001), Goodman (2004),

Leake et al. (2005), Leake and Arber (2006), Tu and Song (2013), and Airapetian et al. (2014) developed MHD models of Alfvén-wave heating for a partially-ionized plasma in a solar or stellar chromosphere. A key component of those types of models is the inclusion and self-consistent calculation of the anisotropic electrical conductivity tensor.

To describe the role of ambipolar diffusion and the Hall effect in the chromosphere of a cool evolved star we need to know how the thermodynamic parameters of the atmosphere vary with height. Those parameters are provided by semi-empirical models of the chromosphere, complemented by the magnetic-field profiles (see Sect. 5.2.1). In the following, as an example we focus on McMurry's semi-empirical model for α Tau. We also consider that the chromosphere has a longitudinal but vertically diverging magnetic field, $B_z = B_0\ (R_\star/r)^2$, where $B_0 = 20\,\mathrm{G}$. The left panel of Fig. 5.3 shows the radial profile of the chromospheric temperature, electron number density, n_e, and the neutral density, n_{H}; the right panel presents the radial profile of the neutral fraction, $n_{\mathrm{H}}/n_{\mathrm{tot}}$. The neutral fraction varies from nearly 100% in the photosphere to about 30% at the top of the chromosphere. The magnetic pressure appears to be dominant with respect to the plasma pressure at heights greater than 0.05 $R_\star$. Thus, most of the chromosphere of a cool giant is a magnetized and weakly-ionized plasma.

In a partially ionized plasma consisting of electrons, protons and neutral particles, the motions of charged particles are strongly affected by electron–neutral and ion–neutral collisions. In a highly collisional plasma of a stellar chromosphere, neutral hydrogen, electrons and protons are efficiently coupled, thus constituting a single fluid if the electron-ion collision frequency is greater than the characteristic frequency of the waves propagating through the medium. This approximation is valid for low-frequency Alfvén waves (less than 1 Hz) in a chromosphere. Electrons and protons gyrate along the magnetic field with characteristic frequencies of $f_{ce} = 2.8 \times 10^6\,B$ for electrons and $f_{ci} = 1.52 \times 10^3\,B$ for protons. For a longitudinal magnetic field of 100 G, such frequencies are in the range 10^5–10^8 Hz.

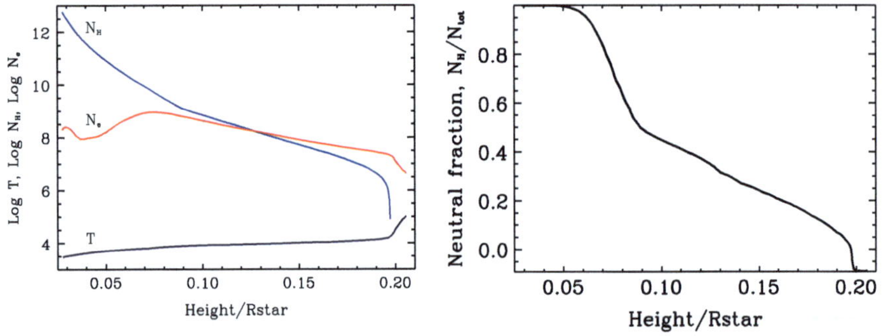

Fig. 5.3 Modelling the chromosphere of α Tau. *Left:* radial profiles of temperature, T, and neutral and electron densities, n_{H} and n_e (in cm^{-3}). *Right:* radial profile of the neutral fraction, $n_{\mathrm{H}}/n_{\mathrm{tot}}$

Unlike fully ionized plasmas, the photosphere and chromosphere of a red giant contain a large fraction of neutral species that collide with electrons and ions at frequencies f_{en} and f_{in}, respectively. Those frequencies are given as

$$f_{en} = 1.95 \times 10^{-10}\, n_{\mathrm{H}}\, \sqrt{T}, \tag{5.14}$$

$$f_{in} = 7.87 \times 10^{-11}\, n_{\mathrm{H}}\, \sqrt{T}, \tag{5.15}$$

and are found to vary within a frequency range of 10^2–10^4 Hz. Consequently, over the entire range of relevant heights in the atmosphere, the electron and ion gyrofrequencies exceed the ion–neutral and electron–neutral collision frequencies. This means that both electrons and ions are magnetized throughout the chromosphere. It also suggests that the Hall effect is negligible in chromospheric conditions, but—according to Goodman (2000)—is important in the lower parts of the solar chromosphere. Mitchner and Kruger (1973) showed that in plasmas where the magnetization of electrons, M_e, and ions, M_i (i.e., the ratio of electron or ion gyrofrequency to the total collision frequency of electrons or ions with neutral particles), becomes greater than 1, the plasma conductivity becomes anisotropic. First, this requires that the Spitzer resistivity, which is parallel to the magnetic field, needs to be modified from its fully-ionized value by adding the frequency of electron–neutral collisions, f_{en}, to the frequency of electron–ion collisions, f_{ei}. Secondly, it also requires including the perpendicular component of the anisotropic electrical resistivity tensor (the Pedersen resistivity), which is described by

$$\eta_{\mathrm{per}} = \frac{[(1+\Gamma)^2 + M_e^2]}{(1+\Gamma)} \eta_{\mathrm{par}} \tag{5.16}$$

$$\eta_{\mathrm{par}} = \frac{m_e(f_{ei} + f_{en})}{n_e\, e^2} \tag{5.17}$$

$$\Gamma = (\frac{n_{\mathrm{H}}}{n_{\mathrm{tot}}})^2\, M_e\, M_i \tag{5.18}$$

$$M_e = \frac{f_{ce}}{f_{ei} + f_{en}}, \quad M_i = \frac{f_{ci}}{f_i}, \tag{5.19}$$

where $f_i = \frac{m_{\mathrm{H}}}{m_i + m_{\mathrm{H}}} f_{in}$ is the reduced ion collision frequency, m_e, m_i and m_{H} are the electron, ion and hydrogen mass respectively, and e is the electron charge.

Figure 5.3 (right panel) and Fig. 5.4 (left panel) show that the photospheric and chromospheric plasma is weakly ionized but strongly magnetized for both electrons and ions, and therefore that $\Gamma \gg 1$ for a longitudinal photospheric magnetic field of 20 G. That further suggests that the Hall effect should be negligible in the atmosphere.

For large Γ, the Pedersen resistivity then can be expressed as

$$\eta_{\mathrm{per}} \propto \Gamma\, \eta_{\mathrm{par}} \propto \frac{B_z^2}{n_{\mathrm{H}}\, \sqrt{T}}. \tag{5.20}$$

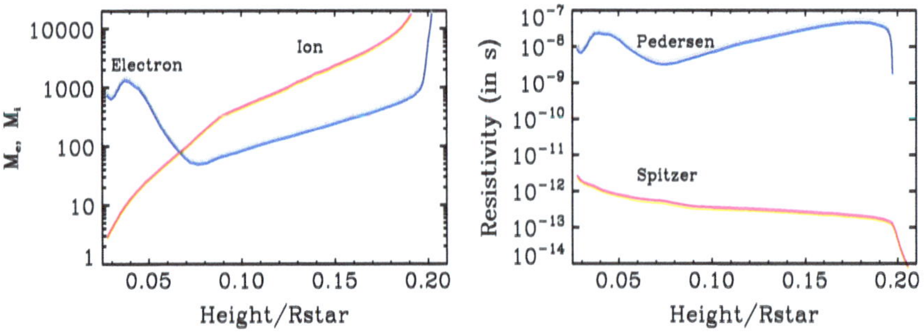

Fig. 5.4 Chromospheric model of α Tau. *Left:* radial profiles of electron and ion magnetizations throughout the chromosphere. *Right:* Spitzer and Pedersen resistivity throughout the chromosphere

The right panel in Fig. 5.4 shows that the Pedersen resistivity is dominant throughout the entire chromosphere of a giant or supergiant. Pedersen resistivity is 4–6 orders of magnitudes greater than the Spitzer resistivity, and should therefore be critical for the heating rates introduced by the dissipation of the electric currents that are induced by the transverse motions of Alfvén waves. This result also means that the less stratified chromospheres of giants and supergiants (because of lower surface gravity) will reduce the density of the chromospheric plasma and therefore increase the significance of the Pedersen resistivity relative to the chromospheres of dwarf stars.

This approach was applied to model chromospheric heating in red giants by Airapetian et al. (2014), who employed a 1.5-D MHD code with a generalized Ohm's law and adopted McMurry's semi-empirical model for α Tau to simulate the propagation of harmonic Alfvén waves at a single frequency of 0.01 mHz. The single fluid, fully non-linear resistive and viscous MHD equations were treated for partially-ionized plasma according to

$$\frac{\partial \rho}{\partial t} + \nabla \cdot (\rho \mathbf{V}) = 0, \tag{5.21}$$

$$\rho \left[\frac{\partial \mathbf{V}}{\partial t} + (\mathbf{V} \cdot \nabla)\, \mathbf{V} \right] = -\nabla P + \rho \mathbf{g} + \mathbf{J} \times \mathbf{B} + \nabla \cdot \mathbf{S} \tag{5.22}$$

$$\frac{\partial \mathbf{B}}{\partial t} = -\nabla \times \mathbf{E}, \tag{5.23}$$

$$\mathbf{E} = -\mathbf{V} \times \mathbf{B} + \left(\eta_{\text{par}} \mathbf{J}_{\parallel} + \eta_{\text{per}} \mathbf{J}_{\perp}\right), \tag{5.24}$$

$$\frac{\partial (\rho E)}{\partial t} + \nabla \cdot (\rho E \mathbf{V}) = -P \nabla \cdot \mathbf{V} + \left(\eta_{\text{par}} \mathbf{J}_{\parallel}^2 + \eta_{\text{per}} \mathbf{J}_{\perp}^2 + \zeta_{ij} S_{ij} - L_{\text{rad}}\right) \tag{5.25}$$

S_{ij} denote the components of the stress tensor $\mathbf{S} = \nu[\zeta_{ij} - (\delta_{ij}\nabla\mathbf{V})/3]$ and $\zeta_{ij} = \frac{1}{2}(\frac{\partial V_i}{\partial x_j} + \frac{\partial V_j}{\partial x_i})$, $\nu = \nu_{\rm nn} + \nu_{\rm in}$ is the viscosity coefficient due to neutral–neutral and ion–neutral collisions (Leake et al. 2013), E is the specific internal energy, given by $E = \frac{P}{\rho(\gamma-1)} + (1 - \xi_n)\frac{X_i}{m_{av}}$, $L_{\rm rad}$ is the total radiative cooling rate; other variables have their usual meaning. The method of solving the continuity, momentum and induction equations in a partially ionized plasma has been described in detail by Arber et al. (2001) and Leake and Arber (2006).

A steady-state flux of Alfvén waves was launched from the photosphere along a vertically diverging flux tube. The initial Poynting flux of Alfvén waves is given as $F_A = 4 \times 10^7\,{\rm ergs\,cm^{-2}\,s^{-1}}$. As waves propagate upwards along the magnetic field lines into the stellar chromosphere, the wave amplitude grows as the density falls with height. Figure 5.5 shows that the amplitude of Alfvén waves increases by a factor 10 at $t = 0.3\ t_A$, where t_A is the Alfvén transit time. Such an amplitude of unresolved turbulence formed by Alfvén waves is consistent with the non-thermal broadening that is observed in UV lines.

One important feature revealed by the simulations of these non-linear transverse waves is their conversion into longitudinal (compressible) waves. The conversion occurs at an altitude $\geq 0.05\ R_\star$, where the sound speed becomes equal to, or greater than, the Alfvén speed (right panel of Fig. 5.6). In the narrow layers where plasma $\beta \sim 1$, non-linear transverse Alfvén wave motions become strongly coupled to compressible wave motions or slow magneto-sonic waves (left panel of Fig. 5.6). The wave mode conversion is revealed by the formation of non-linear density fluctuations at amplitudes as high as 50% of the unperturbed density, starting at $0.05\ R_\star$ and propagating to the upper layers of the chromosphere as presented in the left panel of Fig. 5.6. At that altitude the wave Poynting flux is about $7\times10^5\,{\rm ergs\,cm^{-2}\,s^{-1}}$, and therefore over 70% of the surface energy flux is dissipated or reflected back to the chromosphere. The right panel of Fig. 5.6 shows that the total

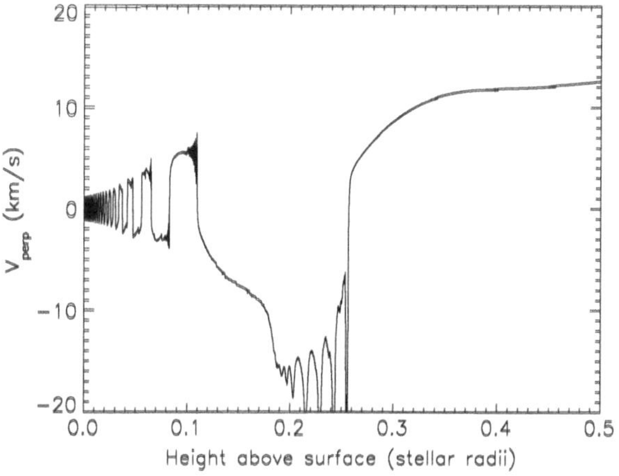

Fig. 5.5 Model output. The radial profile of the Alfvén wave amplitude at $0.3\ t_A$

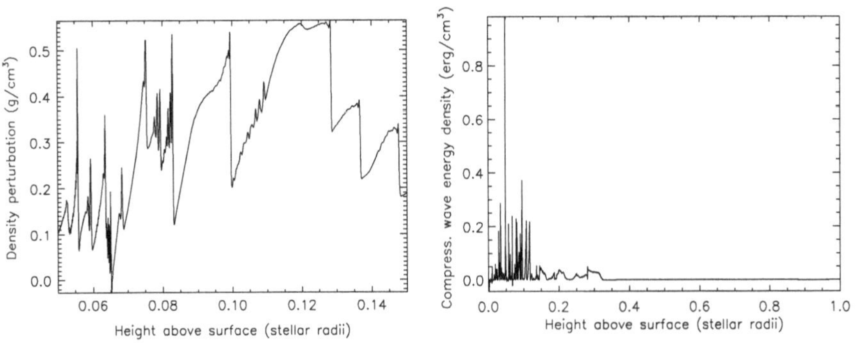

Fig. 5.6 Model outputs. *Left:* radial profile of the density perturbation, normalized to the background plasma density, shows the presence of non-linear slow magneto-sonic waves at $t = 0.3\ t_A$. *Right:* distribution of energy in slow magneto-sonic waves concentrated in narrow layers throughout the chromosphere at $t = 0.3\ t_A$

energy density reaches its first peak at $0.05\ R_\star$, suggesting the formation of non-linear slow magneto-sonic waves at an energy density of $0.25\ \mathrm{ergs\,cm^{-3}}$ in narrow layers distributed in the chromosphere for up to $0.3\ R_\star$. The energy flux associated with compressible waves is $2 \times 10^5\ \mathrm{ergs\,cm^{-2}\,s^{-1}}$, which is $\sim$30% of the flux in the Alfvén waves. Highly non-linear, slow magneto-sonic waves then steepen into shocks that are not resolved in the described simulations. The right panel also shows that the non-linear compressible waves dissipate their energy efficiently into heat in a narrow range of heights, from 0.05–$0.4\ R_\star$. The effect of the dissipation of compressible waves is traced by the sharp drop in the compressible energy density in upward propagating waves.

The wave energy dissipates mostly via viscosity owing to neutral–neutral collisions and resistivity caused by ion–neutral collisions. The effect of mode coupling of the conversion of energy from Alfvén waves into slow magneto-sonic waves has also been observed and described in 1-D MHD simulations of the solar chromosphere by Lau and Siregar (1996), Torkelsson and Boynton (1998), Ofman and Davila (1997), Airapetian et al. (2000) and Suzuki (2013) in simulations of solar and stellar atmospheres and winds. At this point, the Alfvén wave motions become non-linear and induce significant electric fields. The induced perpendicular component of the electric current (with respect to the vertical magnetic field) is then dissipated efficiently by Pedersen resistivity with the volumetric Joule heating rate, $\eta_{\mathrm{per}} J_\perp^2$.

According to the left panel of Fig. 5.1, the McMurry model predicts a strong peak in the Alfvén velocity gradient that forms the reflection point at $\sim 0.2\ R_\star$ for upward propagating Alfvén waves. The model shows that Alfvén waves with critical frequencies less than the critical frequency of $\nu_{\mathrm{crit}} = dV_A/dr = 5 \times 10^{-4}\ \mathrm{Hz}$ get reflected down to the chromosphere and interact with outgoing Alfvén waves, thereby introducing a velocity shear. This suggests that the chromosphere is a low-frequency filter that passes only the higher-frequency waves into the upper atmosphere, where they can be dissipated. The reflected non-linear waves initiate downflows in the chromosphere, which can potentially explain not only non-thermal broadening but also the red-shifts that are observed in UV lines from cool evolved stars.

In a partially ionized plasma, the volumetric viscous heating rate is mostly caused by kinematic neutral–neutral viscosity in the presence of velocity shear, $H_{\text{visc}} = 0.5\zeta_{nn}(\nabla\delta V)^2$. This term becomes important in the energy balance of the chromosphere when waves become strongly non-linear. The left panel of Fig. 5.7 shows that the viscous rate becomes comparable to the Joule heating rate in the mid-chromosphere. However, because the grid resolution is not high enough to allow us to resolve the viscous dissipation scales, viscous effects cannot be estimated consistently.

As non-linear waves propagate upwards, they exert a force commonly known as the ponderomotive force (see also Sect. 5.4.2). It has been shown that it may be responsible for accelerating solar or stellar winds (Ofman and Davila 1997, 1998; Airapetian et al. 2000, 2010).

The model by Airapetian et al. (2014) suggests that non-linear waves can deposit significant momentum and cause the mass loss that is consistent with observation. The right panel of Fig. 5.7 characterizes the momentum deposition in the chromosphere in terms of the mass-loss rate, $\dot{M} = \rho V_r r^2$, throughout the atmosphere at $t = 0.3\ t_A$; V_r is the radial velocity of the plasma, and r is radial distance. The plot shows that at the top of the chromosphere the mass-loss rate attains a constant level of about $10^{18}\ \text{g s}^{-1} \sim 10^{-8}\ \dot{M}_{\odot}\ \text{year}^{-1}$. It also suggests that the filling factor of the open magnetic field is about 0.1%.

Future models should take radiative losses into account and calculate the profiles and fluxes of chromospheric lines formed in the model atmosphere (as discussed in Sect. 5.5). That requires knowledge about the heating rates that balance the radiative losses. In the 1.5-MHD simulations presented above, Airapetian et al. (2014) employed for the first time high-resolution simulations that resolve structures at scales less than 1,500 km. The resolution of their grid allowed them to resolve

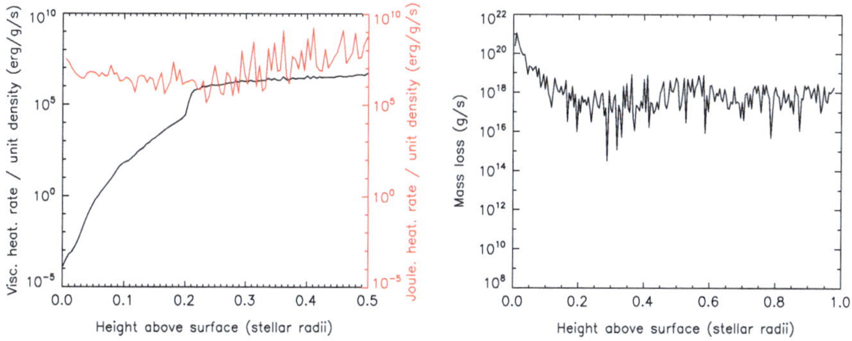

Fig. 5.7 Model outputs. *Left:* viscous (*black/smooth*) vs Joule (*red/jagged*) heating in the chromosphere. *Right:* radial profile of the mass-loss rate at 0.3 t_A

resistive dissipation scales, thereby yielding heating rates that are physically meaningful.

The numerical diffusivity is given by

$$D_{\mathrm{num}} = V_A \Delta^2 / L, \tag{5.26}$$

where Δ is the numerical grid spacing, and L is the characteristic length of the physical structure. D_{num} reaches its largest value when the characteristic length equals the grid size, $V_A \Delta$. The Pedersen diffusivity is,

$$D_{\mathrm{Pedersen}} = c^2 \eta_{\mathrm{per}} / 4\pi, \tag{5.27}$$

and using Eq. 5.20, the ratio of the Pedersen to the numerical diffusivity is therefore proportional to

$$\frac{D_{\mathrm{Pedersen}}}{D_{\mathrm{num}}} \propto \frac{B}{n_{\mathrm{H}} \Delta \sqrt{T}} \,. \tag{5.28}$$

Equation 5.28 suggests that in regions of stronger magnetic fields (i.e. active regions), the ratio of physical to numerical diffusivity, and thus resistivity, increases. Indeed, Fig. 5.8 shows that the ratio of the Pedersen to the numerical resistivity is mostly greater than unity throughout most of the chromosphere of α Tau. It suggests that the calculated heating rates reflect directly the energy input into the plasma and can be used to calculate the radiative output from the chromosphere for direct comparison with observation.

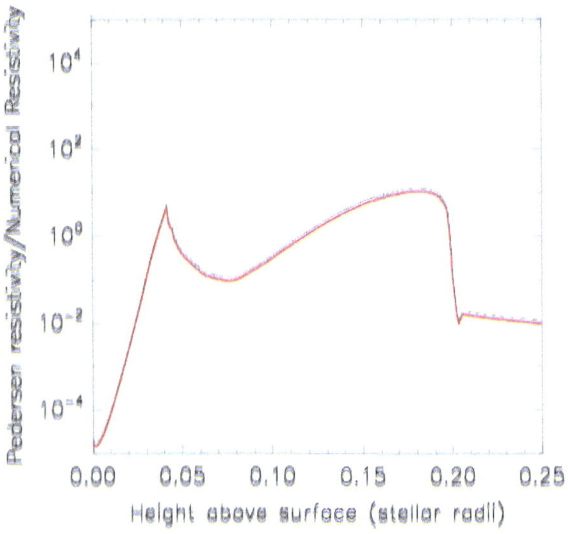

Fig. 5.8 Model output. The vertical profile of the ratio of the Pedersen to the numerical resistivity

5.4.2 *Momentum Deposition by Alfvén Waves: Driving Winds from Cool Evolved Stars*

The general requirements for the driver of winds from cool evolved stars include the condition for their initiation below the sonic point of the star, namely, $R_s = GM_\star/2V_s^2 \sim$ 10–20 $R_\star$, where V_s is the isothermal speed of sound (Holzer et al. 1983). A more refined condition arises from the requirement of initiation and acceleration of winds—for example, in non-coronal giants within the first stellar radius—and its association with the supersonically turbulent and clumpy chromosphere (Carpenter et al. 1995). Radiation pressure on the atmospheric plasma, which is generally accepted as the process driving massive winds from hot stars, is too small to account for what occurs in cool giant stars. Dust-driven effects in wind acceleration are not important in K and early-M giants because there is no evidence of dust formation close to the star (Danchi et al. 1994). Another possible mechanism, of winds driven by acoustic waves, is also not capable of producing the required mass-loss rate, as discussed in Sect. 5.3. The most promising mechanism to date for cool evolved stars requires MHD waves *both* to heat the plasma *and* to accelerate the wind.

The role of MHD waves in initiating and accelerating winds from numerous types of stars, including the Sun, solar-types and cool evolved ones, have been studied extensively since the early 1970s (Hollweg 1973, 1978; Heinemann and Olbert 1980; Hartmann and MacGregor 1980; Ofman and Davila 1998; Airapetian et al. 2000; Suzuki 2007; Cranmer 2008, 2009; Airapetian et al. 2010). The effects of 1-D non-linear Alfvén waves have been studied by Lau and Siregar (1996) and by Boynton and Torkelsson (1996). 2.5-D self-consistent non-linear treatments of Alfvén waves for the solar wind and accelerations from coronal holes have also been performed (Ofman and Davila 1997, 1998). Grappin et al. (2002) modelled a solar wind that was driven by 2.5-D MHD Alfvén waves and included both open and closed magnetic configurations.

The above studies suggest that, as initially small-amplitude (linear) torsional Alfvén waves propagate upwards in a gravitationally-stratified atmosphere along open magnetic field lines, they become non-linear and transfer momentum to the bulk plasma. In general, the acceleration of solar and stellar winds can be driven by the transfer of momentum from the waves to the plasma and cause direct heating of the plasma through the dissipation of MHD waves. The deposition of the momentum of the Alfvén waves occurs through the wave-generated Lorentz force, $\mathbf{J} \times \mathbf{B}$, which is exerted on the plasma (expressed by the momentum equation 5.22). That wave-generated force, or the ponderomotive force (Lundin and Guglielmi 2006), provides plasma acceleration through the gradient of the Alfvén wave pressure, $\frac{1}{\rho}\nabla\left(\frac{B^2}{8\pi}\right)$, and by the magnetic tension force, $\frac{1}{4\pi\rho}(\mathbf{B}\cdot\nabla)\mathbf{B}$. The heating of the wave creates the gradient of the plasma pressure (∇P) in the momentum equation, and can therefore also provide additional acceleration to drive the winds from the Sun and coronal giants. However, this term is not important for accelerating cool winds from non-coronal giants and supergiants, such as the ζ Aur supergiants.

Non-linear coupling of Alfvén waves excites magneto-sonic waves, which can eventually steepen into longitudinal waves that become damped by shock formation. Suzuki and Inutsuka (2005) and Suzuki (2007, 2013) pursued non-linear 1.5-D MHD simulations of Alfvén-wave propagation from solar and stellar photospheres into the corona. While their model does not account for cross-field gradients, they showed that non-linear Alfvén-wave dissipation in the solar atmosphere can explain its thermodynamics through the dissipation of compressible waves produced by mode coupling of non-linear Alfvén waves.

Recent progress in our understanding and modelling of coronal solar and stellar winds driven by MHD waves provides a solid foundation to characterize such winds in the environments of magnetically-active solar-like and evolved luminous stars (Ofman and Davila 1997, 1998; Cranmer 2009; Airapetian et al. 2010; Cranmer and Saar 2011). If the hot, fast, tenuous winds from solar-like stars are driven by a combination of coronal gas pressure and Alfvén wave pressure, the cool massive winds emanating from cool evolved stars are also expected to be powered by Alfvén wave pressure; Airapetian et al. (2000, 2010, 2011) described models of slow massive winds from late-type luminous stars. Future models should apply a single fluid MHD approximation to treat large-scale wind flows from magnetically open fields that extend from the base of the wind to 25 $R_\star$ and beyond. The validity of the MHD approximation is warranted by GHRS spectroscopic observations of wind plasmas that have electron densities of 10^9–$10^{10}\,\mathrm{cm}^{-3}$ and temperatures between 10,000 K and 1 MK. Because the magnetic field at the base of the wind is about 1–10 G, the ratio of the thermal to magnetic pressure in the plasma, or plasma-β, is less that unity at that level.

Airapetian et al. (2000) have described the propagation, in a gravitationally-stratified atmosphere, of non-linear Alfvén waves that are launched from a chromospheric hole at the base of the wind at a single frequency in two wind geometries, and applied it to giants and to supergiants such as α Ori. The results of those simulations suggest the formation of a two-component wind for giant stars from a magnetically open configuration that is dubbed a 'chromospheric hole'. 2-D maps of the radial velocities in that model show the formation of slow (0.1 V_A) and fast (at 0.22 V_A) components of the wind. While the fast component is formed outside the low-density region, the slow wind component emerges from the inner regions of the low-density regions that are about ten times less dense. This model showed that, in order to explain the high mass-loss rate of α Ori, a surface magnetic field of 200 G must be assumed for its chromospherically active regions. Magnetic fields with an average surface distributed field of 1 G have recently been detected in this object (Petit et al. 2013). The finding suggests that the field can reach 200 G if the filling factor of the field is $\sim$0.5%.

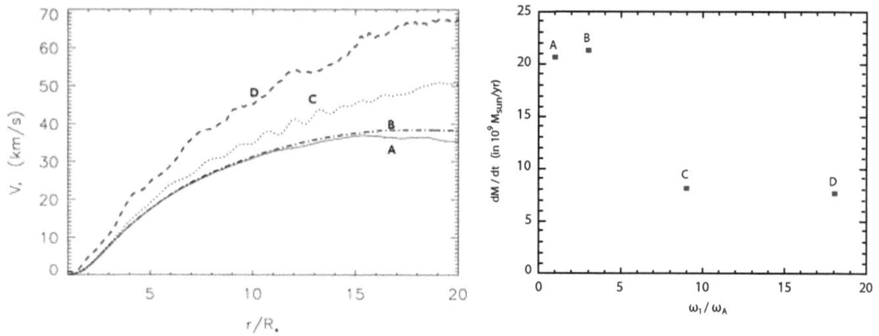

Fig. 5.9 Terminal velocity (*left*) and mass-loss rate (*right*) of a quasi steady-state wind in Models *A*, *B*, *C*, and *D*. ω_1 is the lowest frequency of the corresponding Alfvén wave spectrum (*Reproduced, with kind permission, from Airapetian et al. (2010)*)

Wind models have also been extended by adding a broad-band distribution of Alfvén waves that propagate in the atmospheres of late-type giants (Airapetian et al. 2010). To model the wind from a typical red giant (the example taken was that of α Tau, K5 III), well-constrained input model parameters were adopted (mass, radius, temperature, initial amplitude and spectral range of Alfvén waves launched at the base of the wind), together with output wind parameters inferred from observation (terminal velocity of the wind and mass-loss rate). In order to get a better constraint at the lowest end of the frequency spectrum of the Alfvén waves (ν_1), the parametric dependence of ν_1 on the mass-loss rate and wind speed were assessed by calculating four Alfvén wind models (Models A, B, C, and D) corresponding to various lowest frequencies $\omega_1 = 1, 3, 9$ and $18\,\tau_A^{-1}$ (or wave periods of the Alfvén wave spectrum of 16.2, 5.4, 1.8 and 0.9 days, respectively). At the outer boundary, open (non-reflecting) conditions for MHD waves were imposed in order to reach a quasi steady-state solution. The MHD equations for a fully ionized plasma were solved to calculate the time-averaged terminal velocities and output mass-loss rates of quasi steady-state winds for Models A–D.

Figure 5.9 shows that Model D ($\omega_1/\omega_A = 18$), which corresponds to freely propagating waves, drives faster and less massive winds. The total mass-loss rate from a stellar atmosphere filled by an open magnetic field is proportional to the surface magnetic flux (eq. 11 of Airapetian et al. 2010). That equation suggests that the wind mass-loss rate should vary on timescales of the emergence and evolution of the magnetic flux associated with an open magnetic field, or on the timescale of the rotation period of the star. While the timescale of magnetic field amplification by giant cell convection is of the order of 25 years (Dorch 2004), the rotation period of a typical giant is of the order of 1 year.

The fact that the wind forms in anisotropically distributed 'active regions' on the stellar surface also implies that the generated wind outflows should have anisotropic and clumpy structures. These predictions are consistent with observations of inhomogeneous chromospheres in red giants and supergiant stars including

ζ Aur stars (Eaton 2008; Ohnaka et al. 2013; Ohnaka 2013). Observations by Mullan et al. (1998) and Mészáros et al. (2009) suggest that the mass-loss rate in λ Vel shows time-variability by a factor of six, and by a factor of two in K341 (M15) and L72 (M13), as seen over 18 months of observations.

More specifically, when the total mass-loss rate given by the model is compared with the measured rate for α Tau, the area filling factor of the open magnetic field over the stellar surface (a free model parameter) can be constrained. Those models suggest that winds are initiated in magnetically open structures (chromospheric holes) associated with active regions in cool evolved stars. As new magnetic flux emerges to the stellar surface, it becomes open at a certain height where the kinetic energy of the flow is greater than the magnetic energy. The Alfvén-wave flux generated at the photosphere by convective motions (see Eq. 5.9) is proportional to the surface magnetic flux (Airapetian et al. 2010). The energy flux of the stellar wind is derived from the wave flux, and the mass loss rate of the wind should be scaled by the surface magnetic field.

Results of that kind therefore allow us to provide the framework for sets of physics-based models of winds for cool evolved stars. However, while they provide valuable insights into the dynamics of Alfvén-driven winds in giants, they have not addressed the important aspect of the heating that is associated with Alfvén waves. Moreover, all multi-dimensional MHD simulations of stellar winds performed to date have assumed a fully-ionized plasma, an approximation that is not appropriate for the atmospheres and winds of cool evolved giants (Suzuki 2007; Airapetian et al. 2010).

As discussed in Sect. 5.4.1, the effects of ion–neutral coupling have a major effect on the propagation of Alfvén waves in a stellar chromosphere. It is therefore expected that the wave momentum deposited in the upper atmosphere should be significant in driving cool stellar winds. This new generation of models should be applied to larger sets of non-coronal giants and supergiants in order to assess the generation of mass loss and winds as a function of stellar effective temperature, surface gravity and frequency spectrum of MHD waves.

In conclusion, we should mention a theoretical study by Shukla and Schlickeiser (2003) of the effect of Alfvén-wave propagation through a charged dusty environment. Their model produces dust acceleration through the ponderomotive forces exerted by non-linear waves. It would be interesting to explore its applicability for producing massive dusty winds in AGB stars like late-M Mira variables. The detection of a surface magnetic field in a Mira star (χ Cyg) may suggest the existence of magneto-convective turbulence that is capable of exciting MHD waves (Lobel et al. 2000; Lébre et al. 2014).

5.5 Future Work: Toward Self-Consistent MHD Models of Stellar Atmospheres and Winds

Motivated by the recent progress in multidimensional MHD simulations of solar and stellar chromospheres and winds from luminous late-type stars, (e.g., Martínez-Sykora et al. 2012; Cheung and Cameron 2012; Tu and Song 2013; Airapetian et al. 2014), we can specify four major goals for future self-consistent atmospheric models in ways that can provide efficient direct comparison with observations.

1. The model should resolve physically meaningful scales of energy dissipation. A computationally efficient 1.5-D model is a potentially useful tool for achieving a grid resolution in which the numerical resistivity and viscosity are smaller than their physical values at typical chromospheric parameters. As we move from 2-D to 3-D models, the implementation of such fine grids becomes computationally challenging. In general, because Spitzer resistivity is usually extremely small, even 1-D models encounter problems with the huge number of steps and small time increments. However, the situation becomes attractive in a stellar chromosphere, where Pedersen resistivity is over 4–6 orders of magnitude greater than the Spitzer resistivity. 1-D and 2-D models currently come close to resolving resistive effects; however, the proper inclusion of the viscous dissipation is an even more challenging task, because it requires a resolution that is orders of magnitude more fine. The progress of implementing finer grids will take new physics into consideration, such as the propagation, dissipation, mode conversion and reflection of Alfvén waves in partially-ionized atmospheres.
2. Physically realistic chromospheric models should include the radiative cooling rates from optically thick chromospheric environments in a self-consistent manner. Radiative losses represent a major energy sink in a stellar chromosphere because thermal conduction is negligibly small at temperatures less than a few 0.1 MK. Anderson and Athay (1989) suggested using an 'effectively thin' approximation for a stellar chromosphere, because emission lines of Fe II, Ca II and Mg II are the main contributors to radiative losses under effectively thin conditions; they derived a simple analytical form for the total radiative loss at $T \leq 10{,}000$ K. Goodman and Judge (2012) generalized their expression for total radiative loss in terms of a three-level 'hydrogen' atom with two excited states that is valid for $T < 15{,}000$ K. That expression can provide a good starting point for high-resolution numerical models. The next step will include the coupling between MHD and radiative-transfer codes for a non-LTE atmosphere similar to the features implemented in the 3-D radiative-MHD codes like Bifrost and MURaM, which are used to model the solar atmosphere (Martínez-Sykora et al. 2012). The recent version of the code includes the generalized Ohm's law for both electrons and ions to account for the effects of partial ionization. However, the limitations of 3-D models described in item 1 above do not allow the underlying code to yield physically meaningful heating rates for comparison with observations. For example, with the finest resolution used in the 2-D

MHD Bifrost code, the numerical resistivity is from one to three orders of magnitude greater than the Pedersen resistivity of Martínez-Sykora et al. (2012). That suggests that in the near future 2-D and 3-D radiative-MHD models could provide realistic descriptions of solar and stellar atmospheres.

3. One can generalize the chromospheric models to two and three dimensions that apply the heating in a parameterized but physically meaningful way, coupled to the full radiative-transfer numerical code. Once dissipation rates are characterized in 1.5-D models, one can scale them into the energy equation described by multidimensional models. Such models will consistently describe the propagation, dissipation, mode conversion and reflection of Alfvén waves in partially-ionized atmospheres.
4. In order to study the heating and acceleration of stellar winds consistently with chromospheric heating simulations, future global MHD models should be capable of extending the outer boundary to tens of stellar radii. Those future models should include a unified, fully thermodynamic model of chromospheres of evolved luminous stars that are heated by Alfvén waves and have Alfvén-wave driven winds; models should be given for a significant range of mass, including intermediate ones. That new generation of models will allow one to study the thermodynamics of winds from cool evolved stars, including (but not limited to) the primaries of ζ Aur systems. Such efforts will also allow conditions to be set for future calculations of synthetic spectra. In addition, important efforts should consider the relevance of other kinds of processes, such as magnetic and non-magnetic chromospheric turbulence and waves, especially for low-gravity supergiants. The latter processes have the general ability to lift material closer to the critical point, and so help support the initiation of mass loss (e.g., Schröder and Cuntz 2005; Suzuki 2007, 2013, and references therein). Subsequent ideas and formalisms for the treatment of magnetic and non-magnetic processes for the initiation of mass loss in evolved coronal and non-coronal stars have been forwarded by Cranmer and Saar (2011) while taking stellar magnetic activity into account by extending standard indicators of age, activity and rotation to include the evolution of the filling factors of photospheric magnetic regions.

The ratio of X-ray to bolometric luminosity, L_X/L_{bol}, varies dramatically across the 'dividing line' of Linsky and Haisch (1979). Rapidly rotating giants (earlier than about K3) on the left side of the 'dividing line' show up to six orders of magnitude greater L_X/L_{bol} than do slowly rotating giants in the same temperature range (Ayres 2005). It suggests that the rate of stellar rotation governs the amount of X-ray emission observed in stars across the dividing line. If the magnetic field in evolved stars is generated by a magnetic dynamo, then the rotation rate should scale with the surface magnetic field. Observations seems to suggest that rapidly rotating early-type giants contain the signatures of magnetically-controlled hot coronal plasma and flare activity. Surface magnetic fields have been detected directly in a number of those stars (Konstantinova-Antova et al. 2012). 2.5-D MHD simulations of the emergence of magnetic flux into a partially-ionized solar atmosphere performed by Leake and Linton (2013) suggest that larger magnetic flux emerges faster and reaches a greater height. If applied to stellar environments, that model would suggest that late-type, slowly rotating giants and supergiant stars with weak magnetic fluxes should show signatures of magnetic regions that slowly (i.e., within a few years) emerge into the lower parts of the atmosphere and form compact, closed magnetic loop structures. Correspondingly, rapidly-rotating early-type giants should generate strong magnetic fluxes that are much more buoyant, can reach much greater heights, and can form regions of more extended coronal loops. In those cases the coronal X-ray emission will be determined by the total volume filled by the closed magnetic loops, somewhat like the ones observed in the Sun.

It is fundamental to appreciate that there is an intricate interplay of different processes which are operating on different scales and are responsible for producing the observed phenomena and their signatures. The development of realistic multi-dimensional MHD-wave driven atmosphere or wind models for cool stars is therefore both timely and appropriate.

References

Airapetian, V.S., Ofman, L., Robinson, R.D., et al.: in Cool Stars, Stellar Systems, and the Sun, ed. by R.A. Donahue, J.A. Bookbinder (ASPC **154**, San Francisco, 1998), p. 1569
Airapetian, V.S., Ofman, L., Robinson, R.D., et al.: ApJ **528**, 965 (2000)
Airapetian, V.S., Carpenter, K.G., Ofman, L.: ApJ **723**, 1210 (2010)
Airapetian, V.S., Ofman, L., Sittler, E.C., Jr., et al.: ApJ **728**, 67 (2011)
Airapetian, V.S., Leake, J., Carpenter, K.G.: in Proceedings of the 18th Cambridge Workshop on Cool Stars, Stellar Systems, and the Sun, ed. by G. Van Belle, H. Harris (Proceedings of Lowell Observatory, 2014)
Alfvén, H.: Nature **150**, 405 (1942)
An, C.-H., Suess, S.T., Moore, R.L., et al.: ApJ **350**, 309 (1990)
Anderson, L.S., Athay, R.G.: ApJ **346**, 1010 (1989)
Arber, T.D., Longbottom, A.W., Gerrard, C.L., et al.: JCP **171**, 151 (2001)
Auriére, M., Donati, J.-F., Konstantinova-Antova, R., et al.: A&A **516**, L2 (2010)
Ayres, T.R., Linksy, J.L.: ApJ **200**, 660 (1975)
Ayres, T.R., Simon, T., Stern, R.A., et al.: ApJ **496**, 428 (1998)
Ayres, T.R., Brown, A., Harper, G.M.: ApJ **598**, 610 (2003)
Ayres, T.R.: ApJ **618**, 493 (2005)
Barkhudarov, M.R.: Sol. Phys. **135**, 131 (1991)
Braginskii, S.I.: Transport Processes in a Plasma, **1** (Consultants Bureau, New York, 1965)
Boynton, C.C., Torkelsson, U.: Magnetodynamic Phenomena in the Solar Atmosphere—Prototypes of Stellar Magnetic Activity, IAUC **153**, ed. by Y. Uchida, T. Kosugi, H.S. Hudson (Kluwer, Dordrecht, 1996), p. 467
Brown, A., Deeney, B.D., Ayres, T.R., et al.: ApJS **263**, 107 (1996)
Buchholz, B., Ulmschneider, P., Cuntz, M.: ApJ **494**, 700 (1998)
Carpenter, K.G., Robinson, R.D., Wahlgren, G.M., et al.: ApJ **377**, L45 (1991)
Carpenter, K.G., Robinson, R.D.: ApJ **479**, 970 (1997)
Carpenter, K.G., Robinson, R.D., Wahlgren, G.M., et al.: ApJ **428**, 329 (1994)
Carpenter, K.G., Robinson, R.D., Judge, P.G.: ApJ **444**, 424 (1995)
Carpenter, K.G., Airapetian, V.S.: Cool Stars, Stellar Systems, and the Sun, ed. by E. Stempels (AIPC **1094**, 2009), p. 712

Charbonneau, P., MacGregor, K.B.: ApJ **454**, 901 (1995)
Cheung, M.C.M., Cameron, R.H.: ApJ **750**, 6 (2012)
Chiavassa, A., Haubois, X., Young, J.S., et al.: A&A **515** A12 (2010)
Cranmer, S.R.: ApJ **689**, 316 (2008)
Cranmer, S.R.: ApJ **706**, 824 (2009)
Cranmer, S.R., Ballegooigen, A.A.: ApJS **156**, 265 (2005)
Cranmer, S.R., Saar, S.H.: ApJ **741**, 54 (2011)
Cuntz, M.: ApJ **353**, 255 (1990)
Cuntz, M.: A&A **325**, 709 (1997)
Cuntz, M., Rammacher, W., Ulmschneider, P.: ApJ **432**, 690 (1994)
Cuntz, M., Ulmschneider, P., Musielak, Z.E.: ApJ **493**, L117 (1998)
Cuntz, M., Rammacher, W., Ulmschneider, P., et al.: ApJ **522**, 1053 (1999)
Cuntz, M., Harper, G.M., Bennett, P.D.: A&A **376**, 154 (2001)
Danchi, W.C., Bester, R.W., Degiacomi, C.G., et al.: ApJ **107**, 1469 (1994)
Davila, J.M.: ApJ **317**, 514 (1987)
De Pontieu, B., Haerendel, G.: A&A **338**, 729 (1998)
De Pontieu, B., Martens, P.C.H., Hudson, H.S.: ApJ **558**, 859 (2001)
De Pontieu, B., McIntosh, S.W., Carlsson, M.: Science **318**, 1574 (2007)
Dorch, S.B.F.: A&A **423**, 1101 (2004)
Doyle, J.G., Houdebine, E.R., Mathioudakis, M., et al.: A&A **285**, 233 (1994)
Doyle, J.G., Madjarska, M.S., Roussev, I., et al.: A&A **396**, 255 (2002)
Dupree, A.K., Lobel, A., Young, P.R., et al.: ApJ **622**, 629 (2005)
Eaton, J.A.: AJ **136**, 1964 (2008)
Eaton, J.A., Henry, G.W., Odell, A.P.: ApJ **679**, 1490 (2008)
Eriksson, K., Linsky, J.L., Simon, T.: ApJ **272**, 665 (1983)
Fawzy, D., Ulmschneider, P., Stępień, K., et al.: A&A **386**, 983 (2002)
Fontenla, J.M., Avrett, E.H., Loeser, R.: ApJ **355**, 700 (1990)
Fontenla, J.M., Avrett, E.H., Loeser, R.: ApJ **572**, 636 (2002)
Freytag, B., Steffen, M., Dorch, B.: AN **323**, 213 (2002)
Gary, G.A.: Sol. Phys. **203**, 71 (2001)
Goodman, M.L.: ApJ **533**, 501 (2000)
Goodman, M.L.: A&A **424**, 691 (2004)
Goodman, M.L., Judge, P.G.: ApJ **751**, 75 (2012)
Grappin, R., Léorat, J., Habbal, S.R.: JGR **107**, 1380 (2002)
Gray, D.F.: AJ **135**, 1450 (2008)
Guerreiro, N., Hansteen, V., De Pontieu, B.: ApJ **769**, 47 (2013)
Hansteen, V.: ApJ **402**, 741 (1993)
Hansteen, V.H., Hara, H., De Pontieu, B., et al.: ApJ **718**, 1070 (2010)
Hartmann, L., Avrett, E.H.: ApJ **284**, 238 (1984)
Hartmann, L., MacGregor, K.B.: ApJ **242**, 260 (1980)
Harper, G.M., Brown, A., Bennett, P.D., et al.: ApJ **129**, 1018 (2005)
Harper, G.M.: ApJ **720**, 1767 (2010)
Harper, G.M., O'Riain, N., Ayres, T.R.: MNRAS **428**, 2064 (2013)
Heinemann, M., Olbert, S.: JGR **85**, 1311 (1980)
Hollweg, J.V.: JGR **78**, 3643 (1973)
Hollweg, J.V.: Sol. Phys. **56**, 305 (1978)
Holzer, T.E., Flå, T., Leer, E.: ApJ **275**, 808 (1983)
Holzer, T.E. & MacGregor, K.B.: in Mass Loss from Red Giants, ed. by M. Morris, B. Zuckerman (ASSL **117**, Reidel, Dordrecht, 1985), p. 229
Holzer, T.E.: in Circumstellar matter, ed. by I. Appentzeller, C. Jordan (IAUS **122**, Reidel, Dordrecht, 1987), p. 289
Jess, D.B., Mathioudakis, M., Erdélyi, R., et al.: Science **323**, 1582 (2009)
Jackson, J.D.: Classical Electrodynamics, 3rd ed. (Wiley, New York, 1999)
Judge, P.G.; Tritschler, A.; Chye Low, B.: ApJ Lett. **730**, 4 (2011)

Judge, P.G., Carpenter, K.G.: ApJ **494**, 828 (1998)
Judge, P.G., Cuntz, M.: ApJ **409**, 776 (1993)
Judge, P.G., Stencel, R.E.: ApJ **371**, 357 (1991)
Khodachenko, M.L., Arber, T.D., Rucker, H.O., et al.: A&A **442**, 103 (2004)
Konstantinova-Antova, R.K., Auriére, M., Charbonnel, C., et al.: A&A **524**, 57 (2010)
Konstantinova-Antova, R.K., Auriére, M., Petit, P., et al.: A&A **541**, 44 (2012)
Kopp, R.A., Holzer, T.E.: Sol. Phys. **49**, 43 (1976)
Krasnoselskikh, V., Vekstein, G., Hudson, H.S., et al.: ApJ **724**, 1542 (2010)
Lau, Y.-T., Siregar, E.: ApJ **465**, 451 (1996)
Leake, J.E., Arber, T.D.: A&A **450**, 805 (2006)
Leake, J.E., Arber, T.D., Khodachenko, M.: A&A **442**, 1091 (2005)
Leake, J.E., Linton, M.G.: ApJ **764**, 54 (2013)
Leake, J.E., Lukin, V.S., Linton, M.G.: Physics of Plasmas **20**, 061202 (2013)
Lébre, A., Auriére, M., Fabas, N., et al.: A&A **561**, A85 (2014)
Linsky, J.L., Ayres, T.R.: ApJ **220**, 619 (1978)
Linsky, J.L., Haisch, B.M.: ApJ **229**, L27 (1979)
Lobel, A., Bagnulo, S., Doyle, J.G., et al.: MNRAS **317**, 391 (2000)
Lundin, R., Guglielmi, A.: SSRv **127**, 1 (2006)
MacGregor, K.B., Charbonneau, P.: ApJ **430**, 387 (1994)
Martínez-Sykora, J., De Pontieu, B., Hansteen, V.: ApJ **753**, 161 (2012)
Mathioudakis, M., Jess, D.B., Erdélyi, R.: SSR **175**, 1 (2013)
Matthaeus, W.H., Zank, G.P., Oughton, S., et al.: ApJ **523**, L93 (1999)
McMurry, A.D.: MNRAS **302**, 37 (1999)
Mészáros, Sz., Avrett, E.H., Dupree, A.K.: AJ **138**, 615 (2009)
Mitchner, M. & Kruger, C.H.: Partially Ionized Gases (Wiley, New York, 1973)
Morton, R.J., Verth, G., Jess, D.B., et al.: Nature Comm., **3**, 1315 (2012)
Morton, R.J., Verth, G., Fedun, V., et al.: ApJ **768**, 11 (2013)
Mullan, D.J., Carpenter, K.G., Robinson, R.D.: ApJ **495**, 927 (1998)
Musielak, Z.E., Ulmschneider, P.: A&A **386**, 606 (2002)
Narain, U., Ulmschneider, P.: SSR **54**, 377 (1990)
Narain, U., Ulmschneider, P.: SSR **75**, 453 (1996)
Neilson, H.R., Lester, J.B., Haubois, X.: in 9th Pacific Rim Conference on Stellar Astrophysics, ed. by S. Qain, K. Leung, L. Zhu, S. Kwok, (ASPC **451**, San Francisco, 2011), p. 117
Noyes, R.W, Hartmann, L.W., Baliunas, S.L., et al.: ApJ **279**, 763 (1984)
Ofman, L., Davila, J.M.: ApJ **476**, 357 (1997)
Ofman, L., Davila, J.M.: JGR **103**, 23677 (1998)
Ohnaka, K.: A&A **553**, 8 (2013)
Ohnaka, K., Hofmann, K.H., Schertl, et al.: A&A **555**, 18 (2013)
Osterbrock, D.E.: ApJ **134**, 347 (1961)
Pagano, I., Linsky, J.L., Valenti, J., et al.: A&A **331** (2004)
Pérez Martínez, M.I., Schröder, K.-P., Cuntz, M.: MNRAS **414**, 418 (2011)
Peter, H., Judge, P.G.: ApJ **522**, 1148 (1999)
Peter, H.: A&A **449**, 759 (2006)
Petit, P., Auriére, M., Konstantinova-Antova, R., et al.: Lect. Notes Phys. **857**, 231 (2013)
Piddington, J.H.: MNRAS **116**, 314 (1956)
Pijpers, F.P., Hearn, A.G.: A&A **209**, 198 (1989)
Rabin, D.: ApJ **391**, 832 (1992)
Reimers, D., Hünsch, M., Schmitt, J.H.M.M., et al.: A&A **310**, 813 (1996)
Robinson, R.D., Carpenter, K.G., Brown, A.: ApJ **503**, 396 (1998)
Rosner, R., An, C.-H., Musielak, Z.E., et al.: ApJ **372**, L91 (1991)
Rosner, R., Musielak, Z.E., Cattaneo, F., et al.: ApJ **442**, L25 (1995)
Ruderman, M.S., Berghmans, D., Goossense, M., et al.: A&A **320**, 305 (1997)
Rutten, R.G.M., Schrijver, C.J., Lemmens, A.F.P., et al.: A&A **252**, 203 (1991)
Schrijver, C.J.: A&A **172**, 111 (1987)

Schrijver, C.J.: in Inside the Stars, ed. by W.W. Weiss, A. Baglin (IAUC 137, ASPC **40**, San Francisco, 1993), p. 591
Schrijver, C.J., Pols, O.R.: A&A **278**, 51 (1993)
Schröder, K.-P., Cuntz, M.: ApJ **630**, L73 (2005)
Simon, T., Herbig, G., Boesgaard, A.M.: ApJ **293**, 551 (1985)
Skumanich, A.: ApJ **171**, 565 (1972)
Soler, R., Carbonell, M., Ballester, J.L., et al.: ApJ **767**, 171 (2013)
Song, P., Vasyliunas, V.M.: JGR **116**, A09104 (2011)
Steiner, O.: in Kodai School on Solar Physics, ed. by S.S. Hasan, D.Banerje (AIPC **919**, AIP Publishing, USA, 2007), p. 74
Strassmeier, K.G., Handler, G., Paunzen, E., et al.: A&A **281**, 855 (1994)
Shukla, P.K., Schlickeiser, R.: Phys. of Plasmas **10**, 1523 (2003)
Suzuki, T.K., Inutsuka, S.: ApJ **632**, L49 (2005)
Suzuki, T.K.: ApJ **659**, 1592 (2007)
Suzuki, T.K.: AN **334**, 81(2013)
Tomczyk, S., McIntosh, S.W., Keil, S.L., et al.: Science **317**, 1192 (2007)
Torkelsson, U. & Boynton, G.C.: MNRAS **295**, 55 (1998)
Tsuji, T.: ApJ **540**, L99 (2000)
Tu, J., Song, P.: ApJ **777**, 53 (2013)
Ulmschneider, P.: A&A **222**, 171 (1989)
Ulmschneider, P., Musielak, Z.E., Fawzy, D.E.: A&A **374**, 662 (2001)
Vernazza, J.E., Avrett, E.H., Loeser, R.: ApJS **30**, 1 (1976)
Vernazza, J.E., Avrett, E.H., Loeser, R.: ApJS **45**, 635 (1981)
Velli, M.: A&A **270**, 304 (1993)
Wood, B.E., Linsky, J.L., Ayres, T.R.: ApJ **478**, 745 (1997)

Chapter 6
Optical Interferometry of Giants and Supergiants

Brian Kloppenborg and Gerard van Belle

Abstract Over the last several decades optical interferometers have made substantial gains in ability, evolving from simple two-telescope arrays with 10-m baselines that primarily measured the angular diameters of stars, to four- to six-telescope arrays with 300-m baselines that are capable of imaging objects at high spatial resolution (0.3 milli-arcseconds) and high spectral resolution ($R \sim 30{,}000$). This chapter highlights how optical interferometers have been used during the last three decades to study single and binary systems containing giant and supergiant stars. It reviews diameter measurements and astrometry for single and binary stars, discusses the asymmetric mass-loss processes seen in asymptotic giant-branch stars, shows how resolving stellar disks is helping to solve long-standing problems related to carbon stars, and summarizes some of the state-of-the-art techniques that are now being used to image spots and convective cells on supergiants.

6.1 Diameters and Astrometry of Single and Binary Supergiants

Stars are the building blocks of galaxies, and are also the central engine for their evolution. Detailed knowledge of stellar masses, radii, chemical composition, binary fraction and age are of paramount importance for understanding stellar and galactic structure and evolution. For many decades the only access to those parameters was via application of Russel's binary star theory (Russell 1912a,b), spectroscopic orbit solutions, and lunar occultations (e.g., Evans 1970; White and Feierman 1987). In the last few decades, surveys using optical interferometers have substantially increased the number of stars for which we have observationally-determined radii and accurate orbit solutions.

B. Kloppenborg (✉)
Georgia State University, 25 Park Place South, Suite 605, Atlanta, GA 30302, USA
e-mail: bkloppenborg@chara.gsu.edu

G. van Belle
Lowell Observatory, 1400 West Mars Hill Road, Flagstaff, AZ 86001, USA
e-mail: gerard@lowell.edu

T.B. Ake, E. Griffin (eds.), *Giants of Eclipse: The ζ Aurigae Stars and Other Binary Systems*, Astrophysics and Space Science Library 408,
DOI 10.1007/978-3-319-09198-3_6

In contrast to methods that are statistically based, interferometric observations provide direct evidence of the properties of specific stars against which stellar models can be tested. For example, nearly 200 single main-sequence stars now have observationally-determined radii and temperatures with uncertainties of 3% or less (see Boyajian et al. 2013 and references therein). Interferometry has also re-characterized several exoplanet host stars, resulting in revisions of their habitable zones (e.g., von Braun et al. 2014 and references therein), and has explored stars with solar-like oscillation patterns (Baines et al. 2014).

In the context of giants and supergiants, the radii and effective temperatures of α Her (Benson et al. 1991), α Ori (Dyck et al. 1992), RX Boo, RS Cnc, (Dyck et al. 1995) and an additional 111 giants have been determined with Harvard's Infrared Optical Telescope Array (IOTA) interferometer (Traub et al. 2003). Studies of an additional 143 giants and supergiants, reported in the papers of van Belle et al. (1999) and van Belle et al. (2009), were based on measurements made with the Palomar Testbed Interferometer (PTI; Colavita et al. 1999). Those pioneering studies brought to light some small disagreements between observed and theoretical stellar diameters but which have largely been resolved as distance estimates have improved. However, a bias still exists between diameters determined with interferometric techniques and those that use the infrared flux method, and appears to be limited to stars of <1 milli-arcsecond (mas) in diameter. Casagrande et al. (2014) give details.

In addition to measuring accurate angular diameters, optical interferometers have realized absolute positions with precisions of $\sim$15 mas for single stars and relative positions to $\sim$30 micro-arcseconds for components in binary systems. Binaries that are only slightly resolved by an interferometer will have overlapping interference patterns that can be modelled by one of several analytic functions. If the binary is well resolved but is still within the field of view, a technique called 'Separated Fringe Packets' can be applied; Lane and Muterspaugh (2004) give a comprehensive review of it, and Colavita (1994) provides a demonstration. Separated Fringe Packets is frequently used with the interferometer of Georgia State University's Center for High Angular Resolution Array, CHARA (ten Brummelaar et al. 2006), and achieves orbital parameter and mass estimates to better than 1–5% (Farrington et al. 2010; ten Brummelaar et al. 2014). Binaries that are outside the field of view of the interferometer can be studied using a technique called Dual Beam Narrow-Angle Astrometry (Shao and Colavita 1992) that permits stars within the same isoplanatic patch ($\sim$30–40 arcsec) in the near-IR to be studied (Quirrenbach 1999). This technique involves tracking each star's interference pattern independently, and thus requires two beam combiners and twice the metrology. For stars outside an isoplanatic patch, additional metrology can be installed to achieve 10–20 mas absolute all-sky positioning (Mozurkewich et al. 1988; Shao et al. 1990; Hummel et al. 1994).

One application of these techniques is the study of ζ Aur stars. The diameters of the giants in ζ Aur, 31 Cyg and 32 Cyg were measured interferometrically by Di Bendetto and Ferluga (1990). By combining interferometry, spectroscopy and photometry, Bennett et al. (1996) determined the masses of the components in

ζ Aur to better than 5%, the radii of the components to 2–6%, and the distance to the system with an uncertainty of about 1%.

6.2 Miras and AGB Stars

Asymptotic giant branch (AGB) stars are evolved stars of low to intermediate mass that are nearing the end of their evolution. Their mass losses, which can be as high as $10^{-4} M_{\odot}$ year^{-1} (Matsuura et al. 2009), are caused by pulsations and radiatively driven dusty winds, though the mechanism(s) by which material is lifted from the stellar surface and driven outwards is not understood very well (Woitke 2006; Höfner 2008) despite recent theoretical and observational advances. Improving the understanding of the physical processes involved is important because AGB stars play a crucial role in the chemical enrichment and evolution of galaxies, in that they return a significant amount of material to the interstellar medium.

The high spatial resolution of optical interferometry offers a unique opportunity to study a star from its photosphere out to its immediate circumstellar environment. Although low spatial-resolution images of AGB stars and their surrounding bow shocks often show a high degree of symmetry (see, for instance, Cox et al. 2012), optical interferometers frequently show signs of *asymmetric* mass distribution and clumping within a few stellar radii of the surface.

Low spectral-resolution observations in the H-band and K-band reveal details about a star's photosphere—its diameter, limb darkening, wavelength-dependent opacity and phase-dependent pulsational characteristics (e.g., Quirrenbach et al. 1992; Haniff et al. 1995; van Belle et al. 1996; Perrin et al. 1999; Young et al. 2000; Thompson et al. 2002a,b; Ohnaka 2004; Fedele et al. 2005; Millan-Gabet et al. 2005; Woodruff et al. 2008). Medium and high spectral-resolution observations can probe the spatial extent and presence of any clumping of material in bands of CO, H_2O, and TiO in the so-called 'MOLsphere' (Tsuji 2000) that surrounds both AGB and Mira stars. For example, using the AMBER beam combiner at the *VLTI*, Wittkowski et al. (2011) detected a clump of material that was some 4 mas away from the Mira star R Cnc, yet it contributed nearly 3% of the K-band flux.

Mid-IR observations are well suited for studying the spatial distribution, chemical composition, temperature and mass loss in molecular and dust-forming regions, and also the time-dependence of those regions surrounding AGB stars (e.g., Danchi et al. 1994; Monnier et al. 1997, 2000; Weiner et al. 2006). Observations with the Intensity Stellar Interferometer (ISI) at 11 μm have revealed large changes in density of the material surrounding the Mira star o Ceti that are thought to be signs of inhomogeneities and clumps rather than pulsation-related changes (López et al. 1997). Similarly, simultaneous *VLTI*-MIDI and maser *VLBA* observations of the oxygen-rich Mira variable S Ori indicate increasing mass loss and dust formation near the star's visual minimum, plus spotty SiO maser emissions at some 1.8–2.4 photospheric radii (Wittkowski et al. 2007).

Although optical interferometric data clearly indicate that mass loss from AGB stars is often asymmetric, the cause has yet to be identified. Surface features such as convective cells or magnetic spots will result in an uneven temperature distribution, and will thus cause locally different conditions for dust formation and hence for mass loss as well. Structures of that nature have always been difficult to observe with two- and three-telescope interferometers, but the development of arrays of four or more telescopes, together with recent advances in image reconstruction, now make the testing of such ideas feasible—as mentioned in Sect. 6.4 below. Developments in 3-D hydrodynamical simulations of mass loss may also help explain the apparent discrepancies found between studies at high and at low spatial resolution. Work by Freytag and Höfner (2008) has shown that, although dust formation appears symmetric at large spatial scales, at the smallest spatial scales there are structures that are not radially symmetric. Those may be detected by optical interferometers in the near future.

6.3 Carbon Stars

Carbon stars contribute between 10 and 50% of the carbon enrichment of the interstellar medium (Thronson et al. 1987), so a general knowledge of their fundamental parameters is therefore important for understanding the properties of these core constituents of galactic populations. 'Classical' carbon-star spectra are dominated by features such as C_2, C_2H_2, C_3, CN and HCN (Goebel et al. 1981; Lançon and Wood 2000); the high carbon abundance is attributed to the third dredge-up phenomenon in the transition from AGB stars to carbon stars (Iben and Renzini 1983), when the convective envelope penetrates the inter-shell region between the He- and H-burning shells and leaves the objects with $C/O > 1$.

Until recently, very little fundamental information about carbon stars had been obtained directly. Early work of Dyck et al. (1996a,b) and Bergeat et al. (2001) had resulted in direct determinations of angular diameters and effective temperatures ($T_{\rm eff}$) for a few dozen objects by high-resolution techniques such as interferometry and lunar occultations, though from a wide variety of disparate sources. Those results were consistent with earlier, indirect spectral studies such as that by Cohen (1979), which produced a $T_{\rm eff}$ scale spanning 3240–2230 K as a function of Yamashita spectral class (Yamashita 1972, 1975). More recently, a large homogenous data set of angular sizes for these stars was published by van Belle et al. (2013), thereby allowing a general calibration of the $T_{\rm eff}$ scale to be made. The median effective temperature of the carbon star sample was found to be 2800 ± 270 K, and the median linear radius $360 \pm 100\ R_\odot$.

6.3.1 Effective Temperature Versus $(V_0 - K_0)$

The relationship between the effective temperatures of carbon stars and $(V - K)$, which is a commonly used broad-band colour index, can be compared to the synthetic photometry of Aringer et al. (2009). A plot of $T_{\rm eff}$ versus $(V_0 - K_0)$ is shown in Fig. 6.1 for the objects listed in the catalogue of van Belle et al. (2013). The principal uncertainty in the colour is the V-band variability; its uncertainty has been equated to the scatter in the measurements. Calculating a simple linear fit to the central value of $(V_0 - K_0) = 6$ yields the following relation, and is the one presented in Fig. 6.1:

$$T_{\rm eff} = (2917 \pm 53) - [(V_0 - K_0) - 6] \times (491 \pm 109). \quad (6.1)$$

It gives a standard linear fit with errors in both dimensions, and reduced χ^2_ν of 0.698.

Also plotted in Fig. 6.1 are the corresponding modelled points from fig. 16 of Aringer et al. (2009). By selecting models with $M = 2\,M_\odot, Z = Z_\odot, \log g = 0.0$, and a range of carbon abundances C/O = 1.05, 1.10, 1.40, 2.00, one can compare the predicted dependence of $T_{\rm eff}$ upon $(V_0 - K_0)$ with the observations. There is agreement in the slope of the relationship in the range $(V_0 - K_0) = [4, 5.5]$, although

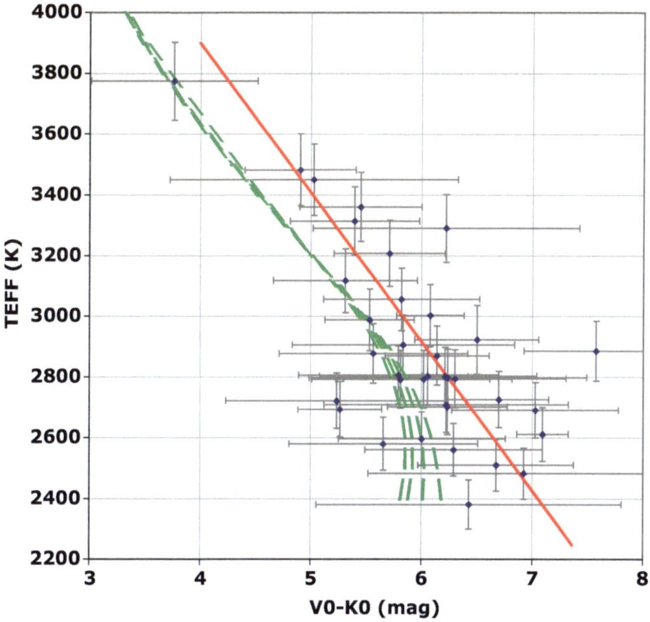

Fig. 6.1 $T_{\rm eff}$ as a function of de-reddened colour $(V_0 - K_0)$. The *solid line* is the linear fit to the data; the *dashed lines* are the $[T_{\rm eff}, (V_0 - K_0)]$ tracks from fig. 16 of Aringer et al. (2009) for models with $M = 2\,M_\odot, Z = 1\,Z_\odot, \log(g) = 0.0$, and C/O = 2.00, 1.40, 1.10, 1.05 (left to right) (*Reproduced, with kind permission, from van Belle et al. (2013)*)

there is an offset in $T_{\rm eff}$ of 200 K; redwards of $(V_0 - K_0) > 5.5$ the models are too low in $T_{\rm eff}$ at a constant colour of $(V_0 - K_0) \sim 6$, a feature not captured by the simple linear fit. However, in a qualitative sense the observational data exhibit an increasing spread in $(V_0 - K_0)$ at a given $T_{\rm eff}$ for $T_{\rm eff} < 3000$ K, which is perhaps indicative of a spread in C/O that is suggested by the synthetic photometry of Aringer et al.

Aringer et al. had noticed the blueward shift of the models relative to observations at a given $T_{\rm eff}$, and had linked it to concerns about the corrections for interstellar reddening that had been applied and to the limitations of non-simultaneous photometry. The former concern is perhaps mitigated by the treatment of reddening described by van Belle et al. The latter is an area where improvements can be made, both from the specific standpoint of simultaneous measurement of V and K photometry, and from the general application of $F_{\rm bol}$ determinations from (spectro-)photometry.

6.3.2 *Asphericity*

There are additional, considerably dense, measurements of angular sizes for a number of the stars in the sample of van Belle et al. (2013) too. Those data sets have been used to demonstrate that some, and potentially all, carbon stars have non-spherical photospheres; two such objects are depicted in Fig. 6.2. There are three possible models to explain those departures from spherical symmetry: (a) the presence of a pronounced stellar spot (a region of increased flux) on a circularly symmetric stellar photosphere, (b) the presence of a disk around a circularly symmetric stellar photosphere, or (c) a photosphere that actually appears to be elliptical.

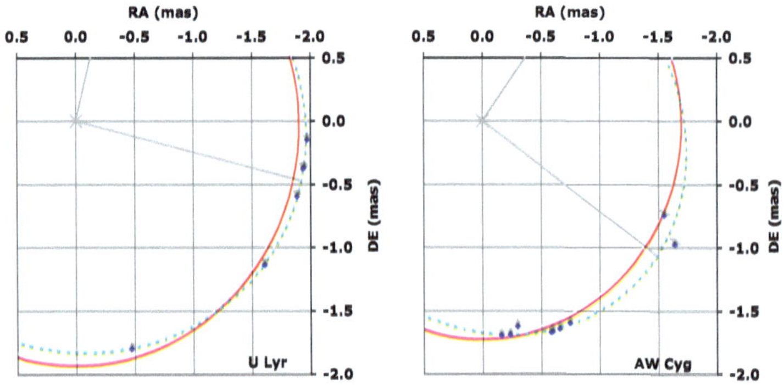

Fig. 6.2 The carbon stars U Lyr (*left*) and AW Cyg (*right*) (after van Belle et al. (2013)). The *solid line* is the best circular fit, and the *dotted line* the best elliptical fit, to the on-sky visibility data from the Palomar Testbed Interferometer. These objects show oblateness ($o_{\rm ab} = a/b - 1$) values of $0.101^{+0.038}_{-0.030}$ and $0.153^{+0.042}_{-0.049}$, respectively

Hot spots seem to be specifically supported by stellar models. Freytag et al. (2008) find that spatial inhomogeneities, induced by huge convection cells, should be significant surface features on AGB stars. Chiavassa et al. (2010b) and Chiavassa et al. (2011a) find that, for red supergiants such as α Ori, significant photocentric excursions of up to ~7.5% of the stellar radius are expected in the visible owing to smaller-scale convection cells; in the H-band, the investigators (Chiavassa et al. 2011b) go on to show that models have larger 'conspicuous spots' due to sub-photospheric convective cells, which have lives of several years. However, van Belle et al. argue that disk or spot structures would not be consistent with the interferometric data.

A truly oblate stellar photosphere can be created simply by rotating a plain spheroid of gas, and in the case of carbon stars the amount of rotation required is surprisingly modest. Furthermore, for carbon stars the degree of oblateness due to rotation may be intrinsically connected to the phenomenon of mass loss. Previous studies by Kahn and West (1985) and Johnson and Jones (1991) established that mass loss for these stars must be aspherical, though without positing an underlying cause. The latter was an empirical study based upon polarization observations, and which have recently been expanded (López and Hiriart 2011a,b). Much of the attention directed towards mass loss by carbon stars has concentrated on the chemistry and grain sizes of dust (Mattsson and Höfner 2011); however, there has also been some discussion in the literature that explores the relationship between rotation and mass-loss rates (e.g., Dorfi and Höfner 1996).

For the carbon star V Hya, a spectroscopic rotation velocity, $v \sin i$, of 13.5 km s^{-1} was acknowledged as "rapid" by Barnbaum et al. (1995), who predicted in their conclusions that the star would "not be spherically symmetric". Such rotation is consistent with the fact that a large fraction (>50%) of planetary nebulæ are aspherical (Zuckerman and Aller 1986), a finding that is potentially connected to the rotation of the central star (Mufson and Liszt 1975; Phillips and Reay 1977; Pascoli 1987).

According to equation A1 of van Belle et al. (2006), the range of oblatenesses found for carbon stars of size $R \sim 360\ R_{\odot}$ corresponds to $v \sin i$ of 9–17 km s^{-1} for $M \sim 2\ M_{\odot}$, and to 7–13.5 km s^{-1} for $M \sim 1.2\ M_{\odot}$. This range of putative $v \sin i$ values actually bridges the region between the somewhat qualitatively distinct regimes of "slow" rotation as defined by Dorfi and Höfner (1996) and the "rapid" rotation as defined by Barnbaum et al. (1995).

Spectroscopic measurements of the rotational velocities of carbon stars are challenging, since those stars have complicated spectra that are principally characterized by a dense set of vibration-rotation transitions of CN and C_2 (Barnbaum et al. 1995). It is quite possible that the rotation rates for carbon stars have not been well measured owing to the difficulty and complexity of the task.

If carbon stars are truly oblate, the parent population could plausibly be evolved rapid rotators, or stars spun up from exoplanet ingestion, as proposed by Soker and Harpaz (2000), Soker (2001) and Carlberg et al. (2009, 2011). Indeed, Soker (2001) argues that ingesting fairly small exoplanets ($M \sim$ 0.01–0.1 $M_{\rm Jupiter}$) within 2 AU could be sufficient to spin up stars as they reach the AGB. If correct, that hypothesis would have been boosted significantly by recent Kepler results which indicate that $\sim$23% of FGK stars have 1–3 $M_{\rm Earth}$ (0.003–0.01 $M_{\rm Jupiter}$) planets within 0.25 AU (Fressin et al. 2013). That result is further extended by indications that the number of planets as a function of orbital period ($\delta N_{\rm planets}/\delta \log P$) remains constant out to 500 days (Dong and Zhu 2012), which implies that $\sim$45% of FGK stars have 1–3 $M_{\rm Earth}$ planets out to 1.2 AU and that additional planets with similar masses at distances between 1.2 and 2.0 AU have yet to be found.

6.4 Supergiants

Supergiant stars represent some of the most extreme situations in stellar astrophysics in terms of mass, composition, nucleosynthesis, radius, temperature and mass loss. The presence of spots and large convective cells on the surfaces of those stars leads to short-term variability (Chiavassa et al. 2011c) and to long-term or stochastic oscillations (Kiss et al. 2006). The presence of surface structure will bias diameter determinations (Chiavassa et al. 2009) and cause non-negligible photocentre shifts that severely degrade the precision of parallax determinations for the largest stars with (say) $R \gtrsim 1$ AU (Eriksson and Lindgren 2007; Pasquato et al. 2011).

The presence of starspots on magnetically-active stars has long been established through their impact on spectral line profiles; the review by Strassmeier (2009), for example, references Doppler images of nearly 100 stars. However, the detectability of convective spots is not well established. Aperture-masking experiments at the William Herschel Telescope in the 1990s detected bright wavelength-dependent asymmetries at visible wavelengths on the M supergiants α Ori, α Sco and α Her (Buscher et al. 1990; Tuthill et al. 1997). Some years later Young et al. (2000) found similar asymmetries at visible wavelengths on α Ori using the COAST interferometer, yet it has a featureless disk in the J-band. This rather baffling result suggests that the bright spots are temporarily unobscured regions seen through a line-blanketed atmosphere. As mentioned above, AGB stars frequently show the presence of non-zero closure phase (e.g., Ragland et al. 2006) that may be due either to surface features or to clumps of material that has recently been lost.

Despite these observational difficulties, optical interferometers have now established that surface features can be detected on supergiants. Two hotspots on α Ori have been detected unambiguously in the H-band by Haubois et al. (2009). Chiavassa et al. (2010a) also detected two spots on the red hypergiant VX Sgr that are thought to be the imprint of giant convective cells.

The advent of optical interferometers with four or more telescopes and advances in image reconstruction techniques are expected to make the detection of starspots a routine operation. Baron et al. (2012) have introduced a new ‘5-D’ image reconstruction program, SQUEEZE, to reconstruct spots on the surfaces of spheres. The software implements compressed sensing theory by representing the image on the most sparse basis possible, thus minimizing the chance of over-fitting the data.

A major difficulty in determining the properties of surface features arises from the way in which model fitting is performed. The traditional method in which models are compared, namely the χ-squared test, is ill-adapted for assessing the relative probabilities of models (Marshall et al. 2006). Two new model-fitting programmes, SIMTOI (Kloppenborg and Baron 2012) and FITNESS (F. Baron et al., unpublished) have more or less abandoned that technique in favour of a Bayesian approach using MultiNest (Feroz and Hobson 2007; Feroz et al. 2008), a Bayesian nested sampling library. By using Bayesian evidence (the average likelihood over the parameter space, or prior), the programmes automatically implement the principle of Occam's razor: a more complex model will only have a higher evidence value if it truly fits the data better. The nested sampling portion of MultiNest allows the minimization process to explore efficiently the multi-modal and/or degenerate probability spaces. In addition to that algorithmic development, SIMTOI uses the Open Compute Language (OpenCL) to leverage high-throughput computing devices, like graphical processing units, to perform calculations 50–100 times faster than the same algorithms do on CPUs. SIMTOI also uses the computer graphics library OpenGL to render models in a fully 3-D, time-dependent environment. Complex geometries such as Roche surfaces, surface features such as spots or convective cells, and non-stellar eclipses such as that of ϵ Aurigae (Chap. 4), can then be modelled with ease.

By using those techniques, and data from the MIRC beam combiner at CHARA, Baron et al. (2014) have produced a definitive detection of spots or convective cells on the red supergiants RS Per and T Per. Similarly, Parks et al. (2012) and Parks et al. (2013) have resolved spots on the surface of the magnetically-active giant λ And.

6.5 To the Future!

Within the last few years optical interferometry has grown from a field concentrating on resolving the diameters of stars and determining orbits of binaries into a technique that can study the surfaces of stars and stellar environments at high spatial and spectral resolution. Classical applications of optical interferometry have resulted in interferometrically-determined radii for more than 1,000 stars. Likewise, long-term astrometric programmes will continue to yield masses, positions and proper motions for the brightest stars. The advent of medium- and high-spectral resolution beam combiners have already yielded a wealth of information concerning the distribution of material seen in the molecular halos that surround Mira and AGB stars, and interferometry has also provided hints that carbon stars may be oblate, despite their relatively modest rotational rates. Optical interferometers have recently achieved conclusive imaging of the surface features on five giants or supergiants. Increasing baseline lengths for optical interferometers are extending the operational reach of new equipment to include main sequence stars, yet are not eclipsing the need to continue to apply such unique capabilities to studies of evolved objects as well.

Within the next decade, planned infrastructure upgrades will increase the sensitivity and resolution limits of existing facilities. New technologies such as integrated optics, photon counting detectors, and more multi-axial beam combiners will make imaging possible at visible and mid-infrared wavelengths, while algorithmic advances using new computing platforms will promote the detailed characterization of complex objects.

This expanding suite of high-resolution instruments on interferometers is allowing the community to take renewed looks at objects that have been studied before but with less advanced instrumentation, and to access targets that were previously too faint or too small for the earlier generations of instrumentation. The ability to study *directly* the fundamental nature of a stellar object is not only important but can be decisive for verifying the indications of indirect measurements (and ϵ Aur is a perfect case in point), for confronting predictions of models, and for making wholly unexpected discoveries.

References

Aringer, B., Girardi, L., Nowotny, W., et al.: A&A **503**, 913 (2009)
Baines, E., Armstrong, T., Schmitt, H. et al.: ApJ **781**, 90 (2014)
Barnbaum, C., Morris, M., Kahane, C.: ApJ **450**, 862 (1995)
Baron, F., Kloppenborg, B., Monnier, J.D.: SPIE **8445**, 1D (2012)
Baron, F., Monnier, J.D., Kiss, L., et al.: ApJ **785**, 46 (2014)
Bennett, P., Harper, G., Brown, A., et al.: AJ **471**, 454 (1996)
Benson, J., Hutter, D., Johnston, K. et al.: AJ **102**, 2091 (1991)
Bergeat, J., Knapik, A., Rutily, B.: A&A **369**, 178 (2001)
Boyajian, T., von Braun, K., van Belle, G., et al.: AJ **147**, 47 (2013)
Buscher, D., Baldwin, J., Warner, P., et al.: MNRAS **245**, 7P (1990)
Carlberg, J.K., Majewski, S.R., Arras, P.: ApJ **700**, 832 (2009)
Carlberg, J.K., Majewski, S.R., Arras, P., et al.: AIPC **1331**, 33 (2011)
Casagrande, L., Portinari, L., Glass, I., et al.: MNRAS **439**, 2060 (2014)
Chiavassa, A., Plez, B., Josselin, E., et al.: A&A **506**, 1351 (2009)
Chiavassa, A., Lacour, S., Millour, R., et al.: A&A **511**, 51 (2010a)
Chiavassa, A., Haubois, X., Young, J., et al.: A&A **515**, A12 (2010b)
Chiavassa, A., Pasquato, E., Jorissen, A., et al.: ASPC **445**, 169 (2011a)
Chiavassa, A., Pasquato, E., Jorissen, A., et al.: A&A **528**, 120 (2011b)
Chiavassa, A., Freytag, B., Masseron, T.: A&A **535**, A22 (2011c)
Cohen, M: MNRAS **186** 837 (1979)
Colavita, M.: A&A **283**, 1027 (1994)
Colavita, M., Wallace, J., Hines, B., et al.: AJ **510**, 505 (1999)
Cox, N.L.J., Kerschbaum, F., van Marle, A.-J., et al.: A&A **537**, A35 (2012)
Danchi, W., Bester, M., Degiacomi, C., et al.: AJ **107**, 1469 (1994)
Di Bendetto, G., Ferluga, S.: A&A **236**, 449 (1990)
Dong, S., Zhu, Z.: arXiv1212.4853 (2012)
Dorfi, E.A., Höfner, S.: A&A **313**, 605 (1996)
Dyck, H.M., Benson, J.A., Ridgway, S.T., et al.: AJ **104**, 1982 (1992)
Dyck, H.M., Benson, J.A., Carleton, N.P., et al.: AJ **109**, 378 (1995)

Dyck, H.M., Benson, J.A., van Belle, G.T., et al.: AJ **111**, 1705 (1996a)
Dyck, H.M., van Belle, G.T., Benson, J.A.: AJ **112**, 294 (1996b)
Eriksson, U., Lindgren, L.: A&A **476**, 1389 (2007)
Evans, D.S.: AJ **75**, 589 (1970)
Farrington, C., ten Brummelaar, T., Mason, B., et al.: AJ **139**, 2308 (2010)
Fedele, D., Wittkowski, M., Paresce, F., et al.: A&A **431**, 1019 (2005)
Feroz, F., Hobson, M.: arXiv 0704.3704 (2007)
Feroz, F., Hobson, M., Bridges, M.: arXiv 0809.3437 (2008)
Fressin, F., Torres, G., Charbonneau, D., et al.: arXiv 1301.0842 (2013)
Freytag, B., Höfner, S.: A&A **483**, 57 (2008)
Goebel, J.H., Bregman, J.D., Witteborn, F.C., et al.: ApJ **246**, 455 (1981)
Haniff, C., Scholz, M., Tuthill, P.: MNRAS **276**, 640 (1995)
Haubois, X., Perrin, G., Lacour, S., et al.: A&A **508**, 923 (2009)
Höfner, S.: A&A **491**, L1 (2008)
Hummel, C., Mozurkewich, D., Elias, N., et al.: AJ **108** 326 (1994)
Iben, I., Renzini, A.: ARA&A **21**, 271 (1983)
Johnson, J.J., Jones, T.J., AJ **101**, 1735 (1991)
Kahn, F.D., West, K.A.: MNRAS **212**, 837 (1985)
Kiss, L., Szabo, G., Bedding, T.: MNRAS **372**, 1721 (2006)
Kloppenborg, B., Baron, F.: https://github.com/bkloppenborg/simtoi (2012)
Lançon, A., Wood, P.R.: A&AS **146**, 217 (2000)
Lane, B., Muterspaugh, M.: ApJ **601**, 1129 (2004)
López, B., Danchi, W., Bester, M., et al.: ApJ **488**, 807 (1997)
López, J.M., Hiriart, D.: AJ **142**, 11 (2011a)
López, J.M., Hiriart, D.: RMxAA **47**, 63 (2011b)
Marshall, P., Rajgru, N., Slosar, A.: Phys. Rev. D **73**, 067302 (2006)
Matsuura, M., Barlow, M., Zijlstra, A., et al.: MNRAS **396**, 918 (2009)
Mattsson, L., Höfner, S.: A&A **533A**, 42 (2011)
Millan-Gabet, R., Pedretti, E., Monnier, J.D., et al.: ApJ **620**, 961 (2005)
Monnier, J.D., Bester, M., Danchi, W., et al.: ApJ **481**, 420 (1997)
Monnier, J.D., Danchi W., Hale, D., et al.: ApJ **543**, 861 (2000)
Mozurkewich, D., Hutter, D., Johnston, K., et al.: AJ, **95** 1269 (1988)
Mufson, S.L., Liszt, H.S.: ApJ **202**, 183 (1975)
Ohnaka, K.: A&A **424**, 1011 (2004)
Parks, J.R., White, R.J., Schaefer, G.H., et al.: in Cool Stars, Stellar Systems, and the Sun, ed. by C.M. Johns-Krull, M.K. Browning, A.A. West (ASPC **448**, San Francisco, 2012), p. 1217
Parks, J.R., White, R.J., Plavchan, P., et al.: AAS **221**, abstr. 137.02 (2013)
Pascoli, G.: A&A **180**, 191 (1987)
Pasquato, E., Pourbaix, D., Jorissen, A.: A&A **532**, 13 (2011)
Perrin G., Coudé du Foresto, V., Ridgway, S., et al.: A&A **345**, 221 (1999)
Phillips, J.P., Reay, N.K.: A&A **59**, 91 (1977)
Quirrenbach, A.: in Proc. 1999 Michelson Summer School, ed. by P. Lawson (1999)
Quirrenbach, A., Mozurkewich, D., Armstrong, J., et al.: A&A **259**, L19 (1992)
Ragland, S., Traub, W., Berger, J.-P., et al.: ApJ **652**, 650 (2006)
Russell, H.N.: ApJ **35**, 315 (1912a)
Russell, H.N.: ApJ **36**, 54 (1912b)
Shao, M., Colavita, M.: A&A **262** 353 (1992)
Shao, M., Colavita, M., Hines, B., et al.: AJ **100** 1701 (1990)
Soker, N.: MNRAS **324**, 699 (2001)
Soker, N., Harpaz, J.: MNRAS **317**, 861 (2000)
Strassmeier, K.: A&AR **17**, 251 (2009)
ten Brummelaar, T., McAlister, H., Ridgway, S., et al.: AJ **628**, 453 (2006)
ten Brummelaar, T., Farrington, C., Mason, B., et al.: AAS **213**, 155.14 (2014)
Thompson, R., Creech-Eakman, M., Akeson, R.: ApJ **570**, 373 (2002a)

Thompson, R., Creech-Eakman, M., van Belle, G.: ApJ **577**, 447 (2002b)
Thronson, H.A., Latter, W.B., Black, J.H., et al.: ApJ **322**, 770 (1987)
Traub, W., Ahearn, A., Carleton, N.: SPIE **4838** 45 (2003)
Tsuji, T.: ApJ **540**, L99 (2000)
Tuthill, P., Haniff, C., Baldwin, J.: MNRAS **285**, 529 (1997)
van Belle, G.T., Dyck, H., Benson, J., et al.: AJ **112**, 2147 (1996)
van Belle, G.T., Lane, B., Thompson, R., et al.: AJ **117** 521 (1999)
van Belle, G.T., Ciardi, D.R., ten Brummelaar, T., et al: ApJ **637**, 494 (2006)
van Belle, G.T., Creech-Eakman, M., Hart, A.: MNRAS **394** 1925 (2009)
van Belle, G.T., Paladini, C., Aringer, B., et al.: ApJ **775**, 45 (2013)
von Braun, K., Boyajian, T., van Belle, G.T., et al.: MNRAS **438** 2413 (2014)
Weiner, J., Tatebe, K., Hale, D., et al.: ApJ **636**, 1067 (2006)
White, N., Feierman, B.: AJ **94**, 751 (1987)
Wittkowski, M., Boboltz, D., Ohnaka, K., et al.: A&A **470**, 191 (2007)
Wittkowski, M., Boboltz, D., Ireland, M., et al.: A&A **532**, L7 (2011)
Woitke, P.: A&A **460**, 9 (2006)
Woodruff, H., Tuthill, P, Monnier, J.D., et al.: ApJ **673**, 418 (2008)
Young, S., Baldwin, J., Boysen, R., et al.: MNRAS **315**, 635 (2000)
Yamashita, Y.: AnTok **13**, 169 (1972)
Yamashita, Y.: AnTok **15**, 47 (1975)
Zuckerman, B., Aller, L.H.: ApJ **301**, 772 (1986)

Chapter 7
Asteroseismology of Eclipsing Binary Stars

Daniel Huber

Abstract Eclipsing binaries have long served as benchmark systems to measure fundamental properties of stars. In recent decades, asteroseismology—the study of stellar pulsations—has emerged as a powerful new tool to study the structure and evolution of stars across the H–R diagram. Pulsating stars in eclipsing binary systems are particularly valuable since fundamental properties such as radii and masses can then be determined using two independent techniques. Furthermore, independently measured properties from binary orbits can be used to improve asteroseismic modelling for pulsating stars in which mode identifications are not straightforward. This chapter provides a review of asteroseismic detections in eclipsing binary stars, with a focus on space-based missions such as *CoRoT* and *Kepler* and empirical tests of asteroseismic scaling relations for stochastic ('solar-like') oscillations.

7.1 Introduction

Asteroseismology has undergone a revolution in recent decades. Driven by multi-site ground-based observing campaigns and high-precision space-based photometry, the number of stars with detected pulsations has increased dramatically. Pulsation frequencies and amplitudes can now be measured with unprecedented precision. In particular, the photometric data provided by the *CoRoT* and *Kepler* space telescopes have enabled asteroseismology to be applied to stars throughout the H–R diagram (e.g., Gilliland et al. 2010; Michel and Baglin 2012; Chaplin and Miglio 2013).

Because some space-based telescopes have relatively large apertures, the majority of stars with high-quality asteroseismic detections are relatively faint, and tend to lack independent observational constraints from classical methods such as astrometry or long-baseline interferometry. Combining independent observations with asteroseismology is crucial for advancing the theoretical modelling of observed

D. Huber (✉)
NASA Ames Research Center, Moffett Field, CA 94035, USA
SETI Institute, 189 Bernardo Avenue, Mountain View, CA 94043, USA
e-mail: dhuber@seti.org

T.B. Ake, E. Griffin (eds.), *Giants of Eclipse: The ζ Aurigae Stars and Other Binary Systems*, Astrophysics and Space Science Library 408,
DOI 10.1007/978-3-319-09198-3_7

oscillation frequencies, and for the validation of asteroseismic relations in order to derive fundamental stellar properties. The full potential of asteroseismology can therefore only be realized if such observations are combined with model-independent constraints on properties such as temperatures, radii and masses.

Eclipsing binaries are classical benchmark systems for determining fundamental properties of stars from first principles. As with measuring oscillation frequencies, the observation of photometric eclipses and spectroscopic RVs can be carried out for relatively faint systems as long as the spectral lines of the component stars can successfully be disentangled. Furthermore, eclipses and pulsations can be measured using the same data. Thus, asteroseismology of the components in eclipsing binary systems has far-reaching potential for improving our understanding of stellar structure and evolution.

7.2 Principles of Asteroseismology

This section provides a brief introduction to the basic principles of asteroseismology. For a more thorough discussion, the reader is referred to Christensen-Dalsgaard (2003), Aerts et al. (2010), or Handler (2013).

7.2.1 Types of Pulsation Modes

Single pulsation modes in spherically symmetric objects (i.e. in the absence of rapid rotation) can be described by the quantum numbers l, m and n. The spherical degree l corresponds to the total number of node lines on the surface; the azimuthal order $|m|$ denotes the number of node lines that cross the equator. The azimuthal order takes values ranging from $-l$ to l (so there are $2l+1$ modes for each degree l), and is important for rotating stars when the degeneracy imposed by spherical symmetry is broken. The special case of radial pulsations is expressed as $l = 0$, and corresponds to a star expanding and contracting as a whole (sometimes also called the 'breathing mode'). Spherical degrees greater than zero are non-radial pulsations; $l = 1$ are dipole, $l = 2$ are quadrupole and $l = 3$ are octupole modes. Figure 7.1 shows examples of pulsation modes for different configurations of l and m. Since stars are observed as point sources, cancellation effects generally prevent the observation of high-degree ($l > 3$) modes. In addition to l and m, oscillation modes are further characterized by the radial order n, the number of nodes along a radius from the surface to the centre of the star.

Stellar pulsations can be separated further into two main types: pressure (p) modes and gravity (g) modes. Pressure modes are acoustic waves propagating through the stellar interior by the compression and decompression of gas, the pressure gradient acting as the restoring force. Gravity modes correspond to pulsations due to the interplay of buoyancy and gravity, with buoyancy acting as

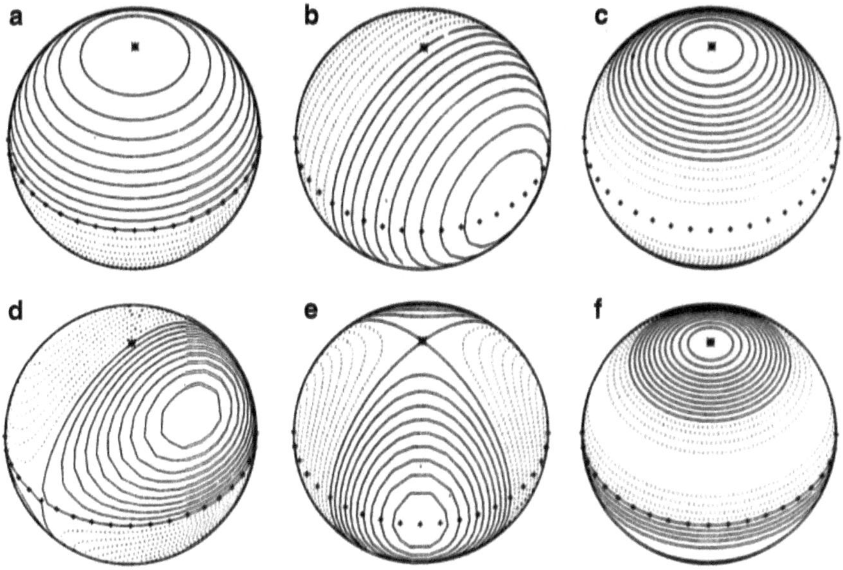

Fig. 7.1 Examples of spherical harmonics used to describe stellar pulsation modes. *Solid lines* show parts of the star moving towards the observer, *dotted lines* parts that move away. The pole of the star is indicated by *asterisks*, the equator by *plusses*. The following cases are shown: (**a**) $l = 1$, $m = 0$; (**b**) $l = 1, m = 1$; (**c**) $l = 2, m = 0$; (**d**) $l = 2, m = 1$; (**e**) $l = 2, m = 2$; (**f**) $l = 3$, $m = 0$ (*Reproduced, with kind permission, from Christensen-Dalsgaard (2003)*)

the restoring force. For g modes, no radial ($l = 0$) modes exist and the radial order n is conventionally counted negative. The propagation zones of p modes and g modes are generally determined by the position of convection zones. Gravity modes are heavily damped in zones where convection is unstable, so in cool stars they are usually confined to the deep interior. Pressure modes, on the other hand, propagate in radiative zones, so are more easily excited to observable amplitudes on the surface. In the case of evolved stars, the p-mode and g-mode cavity can overlap, giving rise to so-called 'mixed modes' (Dziembowski et al. 2001). Such modes are of particular importance for studying interior properties, since they contain contributions from g modes confined to the core but can be observed near the surface. In the case of massive stars with large convective cores, it is more typical to observe g modes.

7.2.2 *Excitation Mechanisms*

Stellar pulsations are excited across a wide range of temperature and evolutionary states in the H–R diagram (see Fig. 7.2). The excitation mechanism driving those oscillations can be broadly divided into two main types: stochastic and coherent.

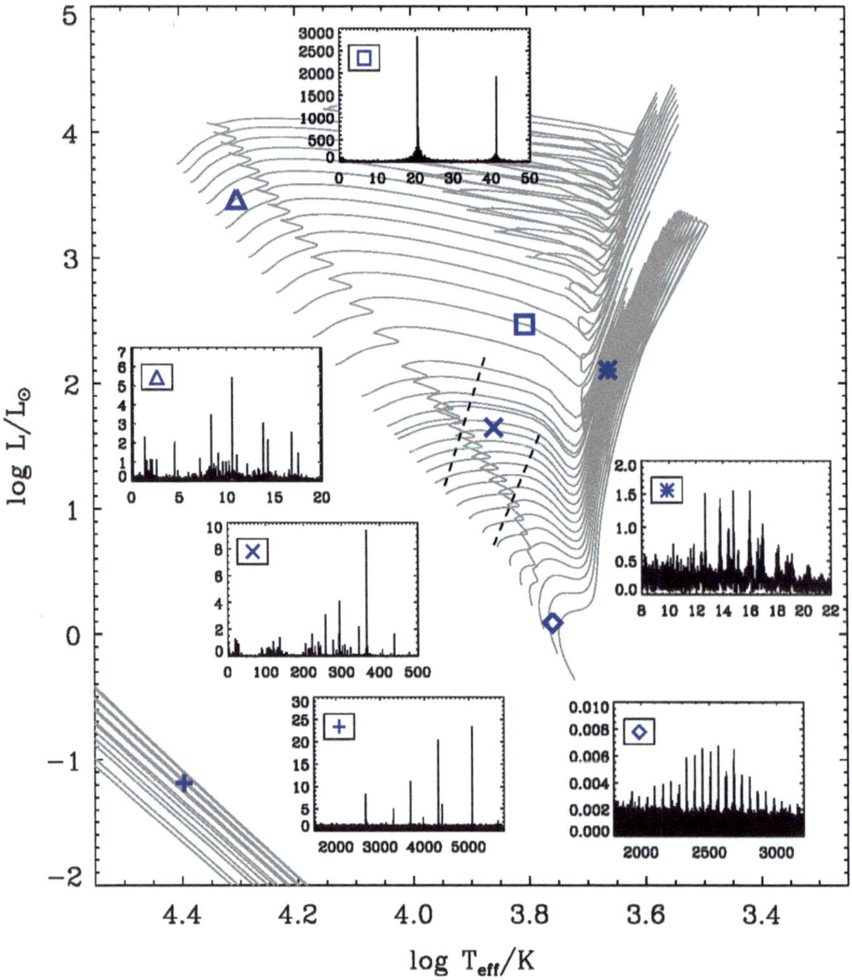

Fig. 7.2 H–R diagram illustrating different types of pulsating stars. *Solid lines* are BaSTI evolutionary tracks (Pietrinferni et al. 2004) and white-dwarf cooling curves (Salaris et al. 2010). *Insets* show amplitude spectra (in units of parts per 1,000) versus μHz for six typical stars based on *Kepler* data: the Sun-like oscillator 16 Cyg B (*diamond*; Metcalfe et al. 2012), the red giant KIC 5707854 (*asterisk*), the hybrid γ Dor-δ Scuti pulsator KIC 11445913 (*cross*; Grigahcène et al. 2010), the hybrid SPB-β Cep pulsator KIC 3240411 (*triangle*; Lehmann et al. 2011), the RR Lyrae star V354 Lyr (*square*; Nemec et al. 2013) and the white dwarf KIC 8626021 (*plus*; Østensen et al. 2011). *Dashed lines* indicate the blue and red edge of the δ Scuti instability strip proposed by Pamyatnykh (2000)

7.2.2.1 Stochastic Oscillations

Oscillations in cool, low-mass stars like the Sun are excited by turbulent convection in the outer layers of the star (e.g. Houdek et al. 1999). The acoustic energy in the convection zone damps or excites oscillation modes stochastically, resulting in mode lifetimes as short as a few days in solar-like stars. Finite mode lifetimes cause the peaks in the oscillation spectrum to be broadened into a Lorentzian shape, so such oscillations typically contain no phase information. Stochastic oscillations are commonly referred to as 'solar-like' oscillations, although they are also found in more evolved stars with convective envelopes, such as red giants.

Stochastic oscillations are typically of high radial order n. Oscillation frequencies $\nu_{n,l}$ of high radial order n and low spherical degree l can be described by the asymptotic theory of stellar oscillations (Vandakurov 1968; Tassoul 1980; Gough 1986), which can be approximated observationally as follows:

$$\nu_{n,l} \approx \Delta\nu(n + \frac{1}{2}l + \epsilon) - \delta\nu_{0l} \, . \tag{7.1}$$

Equation (7.1) predicts that oscillation frequencies follow a series of characteristic spacings. The large frequency separation $\Delta\nu$ is the separation of modes of the same spherical degree l and consecutive radial order n, while modes of the different degree l and same radial order n are expected to be separated by the small frequency separations $\delta\nu_{0l}$.

In the asymptotic theory, the large frequency separation can be shown to be the inverse of twice the sound travel time from the surface to the centre (Ulrich 1986; Christensen-Dalsgaard 2003):

$$\Delta\nu = \left(2\int_0^R \frac{dr}{c}\right)^{-1} , \tag{7.2}$$

where the sound speed c, assuming adiabacity, is given by

$$c = \sqrt{\Gamma_1 P/\rho} \, . \tag{7.3}$$

Here, Γ_1 is an adiabatic exponent, P is the pressure and ρ is the density. The ideal gas law can be expressed as $\rho \propto \mu P/T$, so therefore

$$c \propto \sqrt{T/\mu} \, . \tag{7.4}$$

The sound speed depends on the average internal temperature and chemical composition of the gas. For an ideal gas, basic estimates for the central temperature give $T \propto \mu M/R$ (Kippenhahn and Weigert 1994), and hence

$$\Delta\nu \propto \left(\frac{M}{R^3}\right)^{1/2} . \tag{7.5}$$

The large frequency separation is therefore a direct measure of the mean stellar density.

The power excess of stochastic oscillations has a roughly Gaussian shape, reaching a maximum at a certain frequency. That maximum defines the frequency of maximum power (ν_{max}), and also the maximum amplitude of the oscillation, which is related to the driving and damping of the modes. Brown et al. (1991) suggested that the frequency of maximum power scales with the acoustic cut-off frequency, being the maximum frequency below which an acoustic mode can be reflected (Christensen-Dalsgaard 2003):

$$\nu_{\mathrm{ac}} = \frac{c}{2H_{\mathrm{P}}} . \tag{7.6}$$

Here, H_{P} is the pressure scale height; for an isothermal atmosphere Kippenhahn and Weigert (1994) express it as

$$H_{\mathrm{P}} = \frac{PR^2}{GM\rho} . \tag{7.7}$$

Using the same approximation as above for an ideal gas and assuming that the temperature can be approximated by the effective temperature T_{eff}, we have:

$$\nu_{\mathrm{max}} \propto \nu_{\mathrm{ac}} \propto \frac{M}{R^2\sqrt{T_{\mathrm{eff}}}} . \tag{7.8}$$

Actual frequency spectra are considerably more complex than as described above. For example, frequency separations show variations as a function of radial order depending on the profile of the speed of sound, so can be used to infer details about the interior structure such as the depths of convection zones or stellar ages (e.g., Aerts et al. 2010). However, the simple Eqs. (7.5) and (7.8) readily relate observables to mass and radius, and therefore in principle they offer a straightforward way to calculate those properties. It is important to appreciate that those relations are only approximate, and require careful calibration over a range of evolutionary states. One of the important prospects for asteroseismology of eclipsing binaries is being able to calibrate those scaling relations accurately.

7.2.2.2 Coherent Pulsations

In hotter stars ($T_{\mathrm{eff}} \gtrsim 6500\ K$), pulsations can be driven by temperature-dependent opacity changes that cause the radiation pressure to expand a star continuously past its equilibrium state before it contracts again under the force of gravity. This heat-engine mechanism (also called the κ mechanism) acting in the hydrogen and helium ionization zones is effective in a region of the H–R diagram referred to as the classical instability strip, which includes pulsators such as δ Scuti stars, RR Lyrae stars and Cepheids (see Fig. 7.2). Pulsations driven by the κ mechanism acting in the Fe-group elements drive pulsations in stars hotter than those occurring in the classical instability strip, such as slowly pulsating B (SPB) stars and β Cephei variables. Coherent (also called 'classical') pulsations are phase-stable over long time-spans of stellar evolution, and typically show significantly higher amplitudes than do stochastic oscillations (Fig. 7.2).

Coherent pulsations can also be driven by a heat-engine mechanism which operates at the base of the outer convection zone ('convective blocking': Guzik et al. 2000). That mechanism is the favoured explanation for the g modes observed in γ Dor stars, which border stochastically-driven oscillators and δ Scuti pulsators in the H–R diagram, as well as H-atmosphere white dwarfs (DAV or ZZ Ceti stars). Theoretically the κ mechanism, convective blocking and stochastic driving should be able to excite oscillations simultaneously for stars near the red edge of the instability strip. Recent space-based observations have established that hybrid γ Dor–δ Scuti pulsators are indeed common (Grigahcène et al. 2010), and the first evidence for hybrid coherent-stochastic pulsators has now been found (Belkacem et al. 2009; Antoci et al. 2011).

7.3 The Importance of Eclipsing Binary Stars for Asteroseismology

7.3.1 Asteroseismic Scaling Relations

Quantities derived from detections of stochastic oscillations can be related trivially to fundamental stellar properties by scaling to the observed values of the Sun. For example, Eqs. (7.5) and (7.8) can be rearranged to solve for stellar mass and radius:

$$\frac{M}{M_{\odot}} = \left(\frac{\nu_{\rm max}}{\nu_{\rm max,\odot}}\right)^{3}\left(\frac{\Delta\nu}{\Delta\nu_{\odot}}\right)^{-4}\left(\frac{T_{\rm eff}}{T_{\rm eff,\odot}}\right)^{3/2}, \tag{7.9}$$

$$\frac{R}{R_{\odot}} = \left(\frac{\nu_{\rm max}}{\nu_{\rm max,\odot}}\right)\left(\frac{\Delta\nu}{\Delta\nu_{\odot}}\right)^{-2}\left(\frac{T_{\rm eff}}{T_{\rm eff,\odot}}\right)^{1/2}. \tag{7.10}$$

The large frequency separation $\Delta\nu$ and the frequency of maximum power $\nu_{\rm max}$ can easily be measured from the power spectrum, providing a straightforward (and in principle a model-independent) method for determining fundamental properties of stars. While matching individual oscillation frequencies to stellar models typically yields more precise and detailed information (such as initial helium abundance and age), scaling relations have two important applications. First, owing to the low amplitudes of stochastic oscillations, the S/N ratio is frequently too low to extract reliably a significant number of oscillation modes, so only the average large separation can be determined. Secondly, the measurement of $\Delta\nu$ and $\nu_{\rm max}$ can be performed automatically, thereby enabling the application of asteroseismology to a large number of stars simultaneously. This new era of 'ensemble asteroseismology' has recently been made possible with the launch of space missions such as *CoRoT* and *Kepler*, which provide high-precision photometry for thousands of stars.

By way of illustration, Fig. 7.3 summarizes the detections of stochastic oscillations over the past ~20 years. Starting with the first confirmed detection of oscillations in Procyon by Brown et al. (1991), subsequent ground-based RV campaigns (e.g. by Bouchy and Carrier 2001; Carrier et al. 2001; Kjeldsen et al. 2005; Bedding et al. 2010), together with observations with early space missions such as *MOST* (Matthews 2007; Guenther et al. 2007), yielded detections in about a dozen bright stars. Launched in 2006, *CoRoT* delivered the first high-S/N detections in main-sequence stars (Michel et al. 2009), and led to the breakthrough discovery of non-radial modes in thousands of red giants (De Ridder et al. 2009; Hekker et al. 2009; Mosser et al. 2010). *Kepler*, launched in 2009, continued the revolution of cool-star asteroseismology by populating the low-mass H–R diagram with detections, including dwarfs cooler than the Sun (Chaplin et al. 2011b) and over ten thousand red giants (Stello et al. 2013). The larger number of red giants with detected oscillations is due to a combination of two effects. First, oscillation amplitudes increase with luminosity (Kjeldsen and Bedding 1995), making detection easier at a given apparent magnitude. Secondly, the majority of stars observed by *Kepler* are given 30-min sampling, setting a limit of $\log g \lesssim 3.5$ because less evolved stars oscillate with frequencies above the Nyquist frequency.

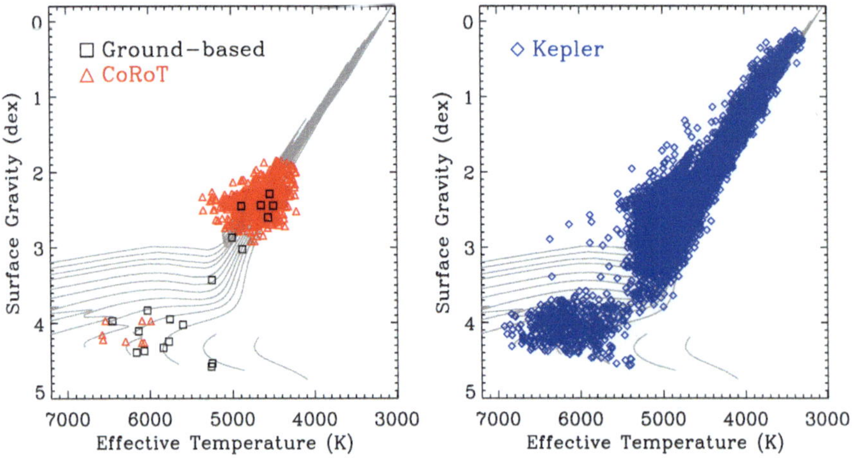

Fig. 7.3 Surface gravity versus effective temperature for stars in which stochastic oscillations have been detected (as of early 2014). *Left:* detections by ground-based campaigns (*squares*) and CoRoT (*triangles*). *Right: Kepler* detections (*diamonds*). Grey lines are solar-metallicity evolutionary tracks to guide the eye. *CoRoT* red-giant detections and *Kepler* detections are taken from the catalogues by Hekker et al. (2009) and Huber et al. (2014), respectively

It is particularly the large number of oscillating giants which drives the need to validate scaling relations. By combining asteroseismic radii and masses with temperatures and metallicities, stellar ages can be determined for thousands of giant stars, thus opening the door to studies of Galactic stellar populations. Indeed, follow-up surveys using multi-object spectrographs such as APOGEE (Mészáros et al. 2013) or Strömgren photometry (Casagrande et al. 2014) have already been dedicated for that purpose. The success of this new era of Galactic archeology relies on our ability to calibrate empirically the asteroseismic scaling relations.

Testing the accuracy of the scaling relations is an active field of research (for reviews, see Belkacem 2012 or Miglio et al. 2013). Theoretical work has shown that both relations typically hold to a few percent (Stello et al. 2009; Belkacem et al. 2011), although deviations of the $\Delta\nu$ scaling relation by up to 2% have been reported for dwarfs with $M/M_{\odot} > 1.2$ (White et al. 2011). Revised scaling relations based on model frequencies (White et al. 2011) and extrapolating the measurement of $\Delta\nu$ to higher radial orders (Mosser et al. 2013) have been proposed, although some doubt about the applicability of the latter revision has been expressed (Hekker et al. 2013). Empirical tests have relied on independently measured properties from *Hipparcos* parallaxes, clusters, and long-baseline interferometry (e.g. Stello et al. 2008; Bedding 2011; Brogaard et al. 2012; Miglio 2012; Miglio et al. 2012; Huber et al. 2012; Silva Aguirre et al. 2012). For unevolved stars ($\log g \gtrsim 3.8$) no empirical evidence of systematic deviations has been determined yet within the observational uncertainties, but for giants a systematic deviation of ~3% in $\Delta\nu$ has been noted for the He-core burning ones (Miglio et al. 2012).

A common limitation is that a separate test of the $\Delta\nu$ and $\nu_{\rm max}$ scaling relation relies on an *independent* knowledge of stellar mass and radius. Such information is only available for stars in binary systems, either when the masses and radii are measured through an astrometric orbit and interferometry, or in doubled-lined spectroscopic eclipsing binaries for which absolute masses and radii can be measured. So far, such a test has only been possible for a three stars: α Cen A, α Cen B, and Procyon A.

7.3.2 Mode Identification and Driving Mechanisms in Intermediate-Mass Stars

Asteroseismology of classical pulsators has been carried out successfully for several decades through ground-based photometric and spectroscopic campaigns (e.g., Breger 2000). Analyses of intermediate- and high-mass stars are plagued by modelling uncertainties such as convective core overshooting and the effects of rotation (e.g., Aerts 2013). Asteroseismology therefore holds great promise for improving our understanding of the evolution of such stars. One serious problem, however, is that classical pulsators often show complex frequency spectra which do not allow mode identification based on simple pattern recognition. In such cases, mode identifications rely on measuring amplitudes in multicolour photometry or spectroscopic line-profile variations, although more recent observations have revealed evidence for systematic structure in the frequency spectra of δ Scuti stars (Breger et al. 2011)[1].

Classical pulsators that occur in eclipsing binary systems offer a possible way to alleviate that problem. Creevey et al. (2011) found that eclipsing binaries constrain fundamental properties of a star as well as, or better than, what is possible for a pulsating δ Scuti star with correct mode identification (Fig. 7.4). Conversely, the correct mode identification can be inferred by comparing solutions that assume different mode identifications, and selecting those solutions which yield the best match to the constraints given by the binary's parameters. Once the mode identification has been secured, interior parameters such as convective core overshoot or mixing length can be constrained further by appealing to the pulsation frequencies.

Eclipsing binary systems may also help us address the long-standing question of the driving mechanisms near the red edge of the instability strip. Stochastic oscillations in δ Scuti and γ Dor stars have been predicted from theory, yet little conclusive observational evidence for hybrid oscillators has been found so far. Pulsating stars with precisely determined fundamental properties may be key to

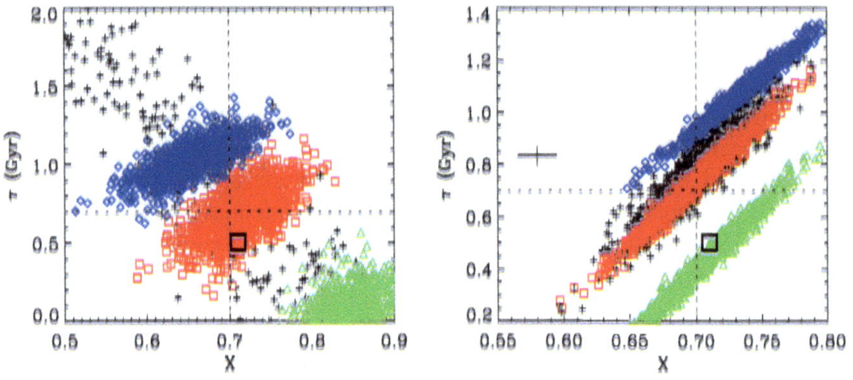

Fig. 7.4 Fitted age versus initial hydrogen abundance for simulated data of a single pulsating δ Scuti star (*left*) and an eclipsing binary including a δ Scuti star (*right*). Symbols denote simulations that include (1) no pulsation mode (*plusses*), (2) a pulsation frequency with a correct mode identification (*squares*), (3) a pulsation frequency with an incorrect spherical degree (*diamonds*), and (4) a pulsation frequency with an incorrect radial order (*triangles*). *Dashed lines* mark the exact input solution, the box is the initial guess values, and the error bar shows the uncertainties in the input for the eclipsing binary solution (*Reproduced, with kind permission, from Creevey et al. (2011)*)

[1] An important exception are rapidly oscillating Ap stars, a class of coherent pulsators that show regularly-spaced high-order p modes (Kurtz 1982).

understanding the interplay between convection and what drives pulsations near the red edge of the classical instability strip.

7.3.3 *Tidally Induced Pulsations and Eccentric Binary Systems*

Multiple star systems offer the possibility to study gravity modes driven by tidal interactions ('dynamical tides': Zahn 1975), which are particularly prominent in eccentric binary systems (e.g., Lai 1997). Tidal interactions can also be invoked to infer properties such as eccentricity, mass and inclination of binary stars, and are of importance for planet formation, for example for migration theories of hot Jupiters through high-eccentricity migration and tidal circularization (Winn et al. 2010).

Observations of dynamical tides have long been hampered owing to the high-precision photometry and continuous coverage required in order to observe the effects over many orbital periods. *Kepler* changed the picture by confirming observationally a new class of eccentric binaries that have tidally-induced brightness variations at periastron passage, also known as 'heartbeat' stars (Thompson et al. 2012 and Sect. 7.4.2). The prototype, KOI-54, consists of two nearly equal-mass A stars in a highly eccentric orbit (Welsh et al. 2011) with tidally-induced pulsations that may be locked into resonance with the binary orbit (Fuller and Lai 2012; Burkart et al. 2012). The discovery of KOI-54 sparked a new era of observational 'tidal asteroseismology'.

While tidal interactions carry a large amount of information about binary systems, some challenges still remain but can be addressed if constraints on parameters are available from eclipse observations. For example, the precise frequencies, amplitudes and phases of tidally-excited oscillations may provide a wealth of information on tidal-damping mechanisms. Independently measured radii and masses of the components then make it possible to model accurately the stellar components and to calibrate theories of tidal dissipation.

7.4 Giant Stars

Eclipsing binaries provide several powerful possibilities for studying the structure and evolution of giant stars, for example through the observation of chromospheric eclipses. But because of their large oscillation amplitudes and the presence of mixed modes, red giant stars also have considerable potential for asteroseismic studies.

7.4.1 Oscillating Giants in Eclipsing Binary Systems

The first detection of an oscillating giant in an eclipsing binary system was presented by Hekker et al. (2010) from *Kepler* data. Figure 7.5, which shows the discovery light curve of KIC 8410637 (TYC 3130-2385-1, $V = 11.3$ mag.), is based on the first 30 days of *Kepler* data; it reveals both a total eclipse (upper panel) and also stochastic oscillations (lower panel). The presence of oscillations during the eclipse suggested that a smaller object was being occulted by a red giant. Since only a single eclipse was observed the orbital period of the system was initially unknown, but constraints from the luminosity and radius ratios indicated that the secondary component is an F-type main-sequence star.

Frandsen et al. (2013) derived a complete orbital solution for KIC 8410637 from nearly 1,000 days of *Kepler* data, an extensive RV campaign and multi-colour ground-based photometry. Figure 7.6 shows the RV solution; combining that with the extended *Kepler* dataset spanning three primary and three secondary eclipses led to a dynamical solution for the system that has an orbital period of 408 days and an eccentricity of 0.6. The radii and masses of the components were determined as $R_{\rm RG} = 10.74 \pm 0.11\, R_{\odot}$ and $M_{\rm RG} = 1.56 \pm 0.03\, M_{\odot}$ for the red-giant primary, and $R_{\rm MS} = 1.57 \pm 0.03\, R_{\odot}$ and $M_{\rm MS} = 1.32 \pm 0.02\, M_{\odot}$ for the main-sequence secondary. The exquisite precision of the absolute radii and masses

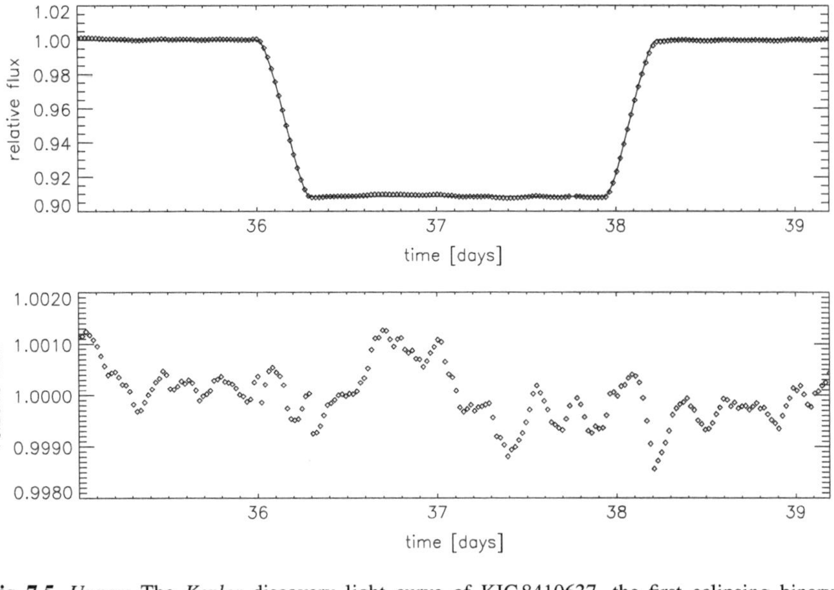

Fig. 7.5 *Upper:* The *Kepler* discovery light curve of KIC 8410637, the first eclipsing binary known to contain an oscillating red-giant component. *Lower:* Light curve after correcting for an eclipse model, revealing the oscillations in the red-giant primary (*Reproduced, with kind permission, from Hekker et al. (2010)*)

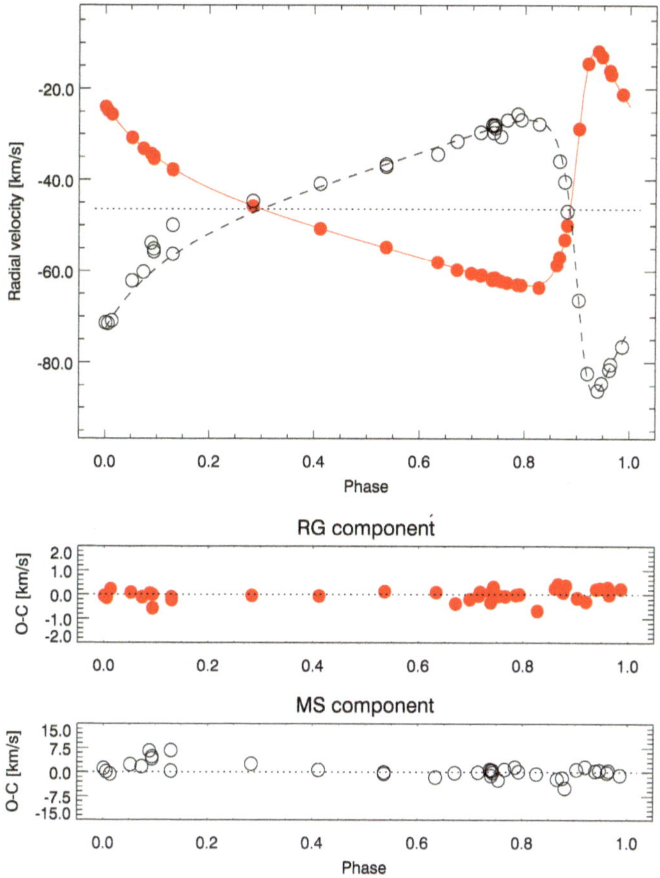

Fig. 7.6 *Upper:* Radial velocities phased with the 408-day orbital period of the red-giant primary (*filled circles*) and the main-sequence secondary (*open circles*) components of KIC 8410637. *Middle and lower:* Residuals of the best-fitting model solution. KIC 8410637 is the first eclipsing binary which has a component that shows stochastic oscillations and also has a measured double-lined spectroscopic orbit (*Reproduced, with kind permission, from Frandsen et al. (2013)*)

make KIC 8410637 an extremely interesting object for testing asteroseismic scaling relations for evolved stars.

Table 7.1 compares the solution by Frandsen et al. (2013) with the values from the scaling relations presented by Hekker et al. (2010). While the values are in reasonable agreement, the uncertainties on the seismic mass and radius are fairly large since the analysis by Hekker et al. was based on only 30 days of *Kepler* data. To test whether the differences are significant, the analysis was repeated using data from Q0-16 (over 1350 days) and the method described by Huber et al. (2009). The resulting seismic values of $\nu_{max} = 46.2 \pm 1.1\,\mu$Hz and $\Delta\nu = 4.634 \pm 0.012\,\mu$Hz are in good agreement with those by Hekker et al. (2010) and also have significantly

Table 7.1 Fundamental properties of the red-giant component of the eclipsing binary system KIC 8410637, derived from an orbital solution and from asteroseismic scaling relations

Parameter	RVs+EB		Asteroseismic scaling relations		
	Frandsen et al.	Hekker et al.	Q1–16[a]	Q1–16+White[b]	Q1–16+Mosser[b]
T_{eff} (K)	4800 ± 80	4650 ± 80	4800 ± 80	4800 ± 80	4800 ± 80
R ($R_\odot$)	10.74 ± 0.11	11.8 ± 0.6	11.58 ± 0.30	11.23 ± 0.29	11.31 ± 0.29
M ($M_\odot$)	1.56 ± 0.03	1.7 ± 0.3	1.83 ± 0.14	1.720 ± 0.13	1.741 ± 0.13
$\log g$ (cgs)	2.569 ± 0.009	—	2.572 ± 0.011	2.572 ± 0.011	2.572 ± 0.011
ρ ($\rho_\odot \times 10^3$)	1.259 ± 0.046[c]	—	1.1765 ± 0.0061	1.2132 ± 0.0063	1.2047 ± 0.0061[c]

[a]For asteroseismic solutions based on Q1-Q16, solar reference values of $\nu_{max,\odot} = 3090 \pm 30\,\mu$Hz and $\Delta\nu_\odot = 135.1 \pm 0.1\,\mu$Hz were adopted (Huber et al. 2011)
[b]Based on scaling-relation corrections proposed by White et al. (2011) and Mosser et al. (2013)
[c]Calculated from the mass and radius

reduced uncertainties (a factor of 2 in mass and radius). Combining the new values with the effective temperature from Frandsen et al. (2013) yields excellent agreement (within 0.003 dex) in $\log g$ (see Table 7.1). However, the density is underestimated by ~7% (1.8 σ, taking the uncertainty in the seismic and dynamical density into account), which results in an overestimate of the radius by ~9% (2.7 σ) and of the mass by ~17% (1.9σ). A difference in density of 7% is a factor of 7 larger than the typical formal uncertainties in the seismic density derived from measurements of $\Delta\nu$ using different methods (Hekker et al. 2012). Table 7.1 also lists estimates of radius and mass using the corrections that have recently been proposed for the $\Delta\nu$ scaling relation by White et al. (2011) and by Mosser et al. (2010). Both corrections reduce the differences to about 4% in density, 5% in radius and 10% in mass.

An important piece of information about KIC 8410637 is knowing whether the primary is a He-core burning red clump star or whether it is still ascending the red-giant branch (RGB). While an asteroseismic determination of the evolutionary state based on gravity-mode period spacings (Bedding et al. 2011; Beck et al. 2011) is still awaited, Frandsen et al. (2013) argue that, from the comparison of the derived temperature with isochrones, a red-clump phase for the primary is the more likely. However, relative corrections to the $\Delta\nu$ scaling relation between red-clump and RGB stars tend to increase the seismic mass and radius (Miglio et al. 2012), and result in even larger differences from the orbital solution. Moreover, as pointed out by Frandsen et al. (2013), the small periastron distance could imply that the system underwent significant mass transfer when the primary reached the tip of the RGB.

Larger samples are clearly required in order to determine whether the differences found for KIC 8410637 could be systematic. Fortunately, *Kepler* offers a goldmine of data for over 15,000 oscillating giant stars that may be used to identify additional systems. Gaulme et al. (2013) cross-matched the *Kepler* eclipsing binary catalogue (Prša et al. 2011; Slawson et al. 2011; Matijevič et al. 2012) with red giants classified in the Kepler Input Catalogue (Brown et al. 2011), and identified

12 new candidate eclipsing-binary systems that also contain oscillating red giants. The red-giant components in the sample span a large range of evolution, making this sample particularly promising for extending tests of the asteroseismic scaling relations. RV follow-ups are currently underway to confirm that these systems are genuine eclipsing binaries with oscillating giants, and to measure their absolute radii and masses.

7.4.2 Oscillating Giants in Eccentric Binary Systems

The first oscillating giants to be discovered in 'heartbeat' systems were presented by Beck et al. (2013), who confirmed the discovery of 18 Kepler systems through RV follow-ups. Figure 7.7 shows five examples of phased heartbeat light curves. The shape, length and amplitude of the light distortion depends on the orientation and inclination of the orbit, as well as on the masses of the components.

Beck et al. (2013) presented a detailed study of KIC5006817, a system with an orbital period of 95 days, an eccentricity of 0.7, and an orbital inclination of 62°. Although the secondary is too faint to be detected in the spectra, the primary is a relatively unevolved RGB star ($R = 5.84\,R_{\odot}$, $M = 1.49\,M_{\odot}$) and shows a high-S/N asteroseismic signal. By appealing to rotationally-split dipole modes (Gizon and Solanki 2003), the asteroseismic analysis yielded an inclination of $77° \pm 9°$ between the stellar axis and the line of sight. The fact that the inclination is larger than that of the orbital inclination, together with a comparison of the orbital period to the rotation period as inferred from rotational splittings, was interpreted as evidence that the system has not yet tidally synchronized.

The analysis of KIC 5006817 yielded several surprising results. First, the data showed an apparent absence of a signal caused by Doppler beaming—a periodic increase and decrease in intensity that is mostly due to the RV shift of the stellar

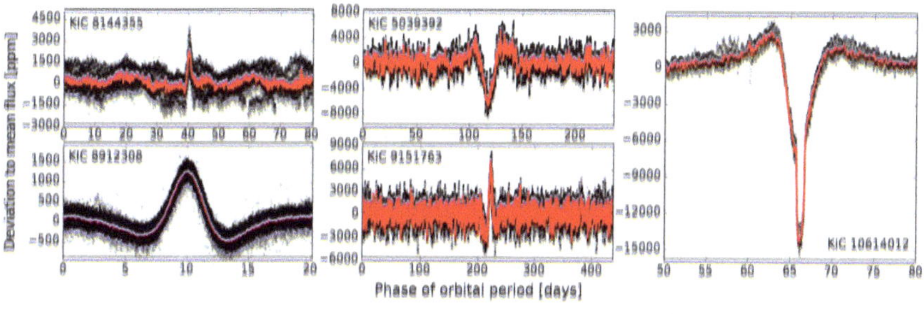

Fig. 7.7 *Kepler* phase curves of five eccentric binary systems containing oscillating red giants, displaying gravitationally induced brightness changes near periastron passage ('heartbeat' stars). The shape and amplitude of the distortion depends on orbital properties such as eccentricity, argument of periastron and inclination, and on the masses of the component stars (*Reproduced, with kind permission, from Beck et al. (2013)*)

spectrum relative to the photometric bandpass. Secondly, the gravity darkening derived from the light-curve model disagreed with empirical and semi-empirical gravity-darkening values. While the former may be related to the difficulty of detrending data when the orbital period is similar to the length of a *Kepler* observing quarter, Beck et al. (2013) concluded that the latter probably implies that a revision of commonly accepted gravity darkening exponents for giants is needed (assuming that the derived properties of the red giant are correct).

While the absence of eclipses and the non-detection of the secondary's spectrum precluded an independent measurement of radii and masses of both components, heartbeat systems such as KIC 5006817 allow insights into the dynamical evolution of eccentric binary systems that contain evolved stars. For example, Beck et al. (2013) found tentative evidence that systems with larger red-giant primary radii have longer orbital periods, suggesting that some of those systems may be the progenitors of cataclysmic variables or subdwarf B stars. Interestingly, the prototype eclipsing binary KIC 8410637 does not show heartbeat events, although its orbital properties are comparable to ones in the sample of Beck et al. which do show them.

7.4.3 Giants in Hierarchical Triple Systems: The Case of HD 181068

An exciting result in the early phases of the *Kepler* mission was the discovery of hierarchical triply-eclipsing triple systems. The first example (KOI-126), presented by Carter et al. (2011), consisted of three low-mass main-sequence stars; a full dynamical solution of all the components was derived by measuring variations in the eclipse timings, thus avoiding the need for RV follow-up observations (a technique that has also been applied to confirm numerous multi-planet systems; see, for example, Fabrycky et al. 2012).

Shortly afterwards Derekas et al. (2011) presented the first discovery of a triply-eclipsing triple system with a red-giant component. Figure 7.8 shows the discovery light curve of HD 181068 (also known as 'Trinity'), which shows long-duration eclipses with an interval of $\sim$23 days interleaved by short-duration eclipses with a interval of $\sim$0.4 days. Follow-up RV and interferometric observations confirmed that the primary is a red giant in an orbital period of 45.5 days and is eclipsed by a pair of main-sequence stars; the latter pair comprise a low-mass binary that eclipses every 0.9 days or so. The short-period eclipses are not seen during primary and secondary eclipse (see Fig. 7.8) because the temperatures (and hence surface brightnesses) of the three components are similar. Subsequent modelling of the variations of the eclipse timings of the outer binary yielded a full dynamical solution. The radii and masses of all three components could be measured to better than 5% (Borkovits et al. 2013).

The HD 181068 system is remarkable for the lack of stochastic oscillations. Figure 7.9 compares a power spectrum of it (after removing all eclipses) to that of

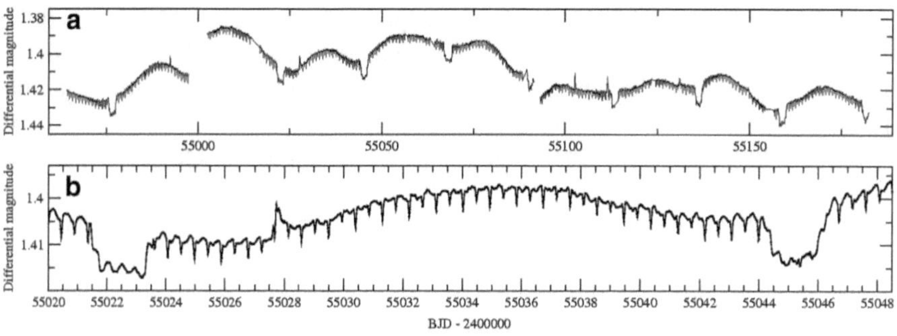

Fig. 7.8 *Upper:* Discovery light curve of HD 181068 ('Trinity'), a triply-eclipsing triple system consisting of a red giant primary and two main-sequence stars. *Lower:* Close-up of one orbital period, showing a primary and secondary eclipse interleaved by ~0.9-day eclipses of the main-sequence binary (*Reproduced, with kind permission, from Derekas et al. (2011)*)

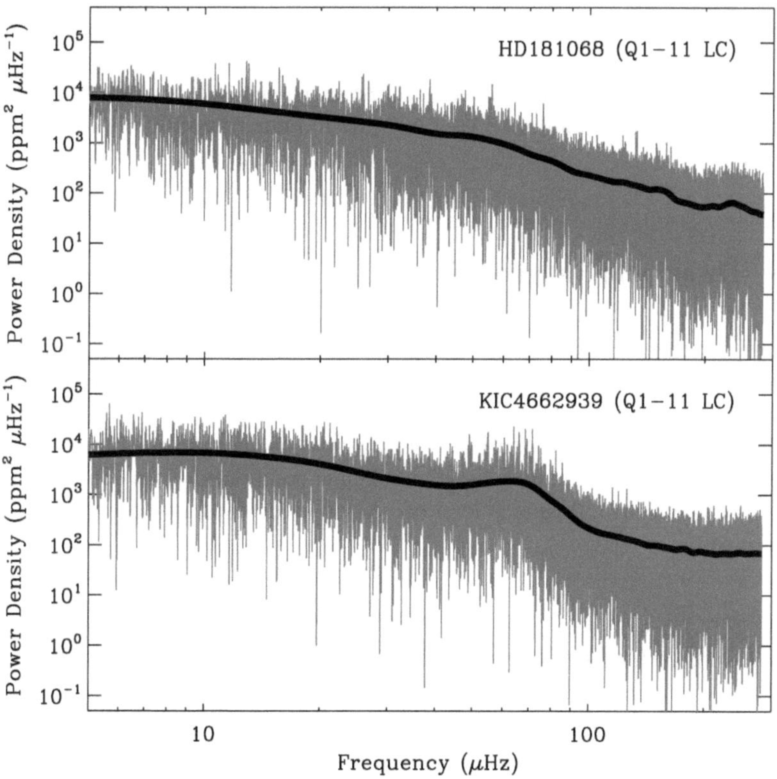

Fig. 7.9 *Upper:* Power spectrum of HD 181068 after removal of the eclipses from the *Kepler* light curve. The *thick line* is a heavily smoothed version of the spectrum. *Lower:* Power spectrum of KIC 4662939, a field red giant whose fundamental properties are similar to those of HD 181068 A. The stochastic oscillations near 70 μHz are absent in the upper panel (*Reproduced, with kind permission, from Fuller et al. (2013)*)

an oscillating field giant whose fundamental properties are nearly identical to those of HD181068 A. It is interesting that the granulation background, which manifests itself as red noise in the power spectrum, is very similar in both stars, while the power excess due to oscillations is completely suppressed in HD181068 A. As speculated by Fuller et al. (2013), the close dwarf components may be responsible for that suppression by tidally synchronizing the rotational frequency of the red giant with the long-period orbit, and causing increased magnetic activity which has been suggested as responsible for suppressing the excitation of stochastic oscillations (Chaplin et al. 2011a). HD181068 A is the first confirmed case of suppressed stochastic oscillations in a binary system, and further evidence that the suppression mechanism is indeed related to binary interactions has been found recently in other candidate eclipsing binaries that contain red-giant components (Gaulme et al. 2014).

Although no stochastic oscillations are observed in HD 181068 A, it does show high-amplitude pulsations at lower frequencies, which can be clearly identified during primary and secondary eclipse (Fig. 7.8). The peaks in the amplitude spectrum are narrow, indicating that they are not stochastically driven, and are linear combinations of the long (ω_1) and short (ω_{23}) orbital frequency with frequencies of $f_1 = 2(\omega_{23} - 2\omega_1)$, $f_2 = 2(\omega_{23} - \omega_1)$, $f_3 = \omega_{23}$ and $f_4 = 2\omega_{23}$. Fuller et al. (2013) demonstrated that those frequencies can be explained by three-body tidal forces that cause the orbital motion of the outer pair to induce pulsations in the red-giant primary.

7.5 Dwarf and Subgiant Stars

Asteroseismology of dwarfs and subgiants in eclipsing binaries has traditionally focused on coherent pulsators such as δ Scuti and γ Dor stars, since they show larger amplitudes than stochastic oscillators, and so eclipses and pulsations can be more easily detected in ground-based observations. At the time of writing, no reports of stochastic oscillations in a dwarf or subgiant component of an eclipsing binary have been published, though at least two *Kepler* detections have conclusively been identified.

7.5.1 Classical Pulsators

Classical pulsations in A and F stars have been detected in ground-based observations in a few dozen detached and semi-detached eclipsing binary systems (see, e.g., Mkrtichian et al. 2004; Rodríguez and Breger 2001, and references therein). However, the short time-base and inferior precision of ground-based photometry often limited the number of reliable pulsation frequencies that could be detected, and only for a small number of those systems could full dynamical orbits be combined with a secure detection of high-amplitude pulsations (e.g., Christiansen et al. 2007).

As with the case of stochastic oscillations, *CoRoT* and *Kepler* have changed that picture dramatically. After the first detections of γ Dor pulsations in eclipsing binaries had been discovered by *CoRoT* (Damiani et al. 2010; Sokolovsky et al. 2010), Maceroni et al. (2009) presented an RV orbit of an eclipsing binary consisting of two B stars that showed additional variations which could be attributed to either self-driven or to tidally-induced pulsations. Those were then followed by firm detections of self-excited γ Dor pulsations (Debosscher et al. 2013; Maceroni et al. 2013), of δ Scuti pulsations (Southworth et al. 2011; Lehmann et al. 2013), and of hybrid γ Dor–δ Scuti pulsations (Hambleton et al. 2013; Maceroni et al. 2014) in double-lined spectroscopic and eclipsing binaries. Those studies illustrated that disentangling the pulsational variability from the variability induced by the binary orbit requires careful iterative techniques (see Fig. 7.10), and also showed up the limitations of current light-curve modelling codes to account for complex reflection effects (e.g., Southworth et al. 2011). Each study constrained the primary and secondary masses, radii and temperatures with uncertainties of a few percent, and yielded a wealth of pulsation frequencies.

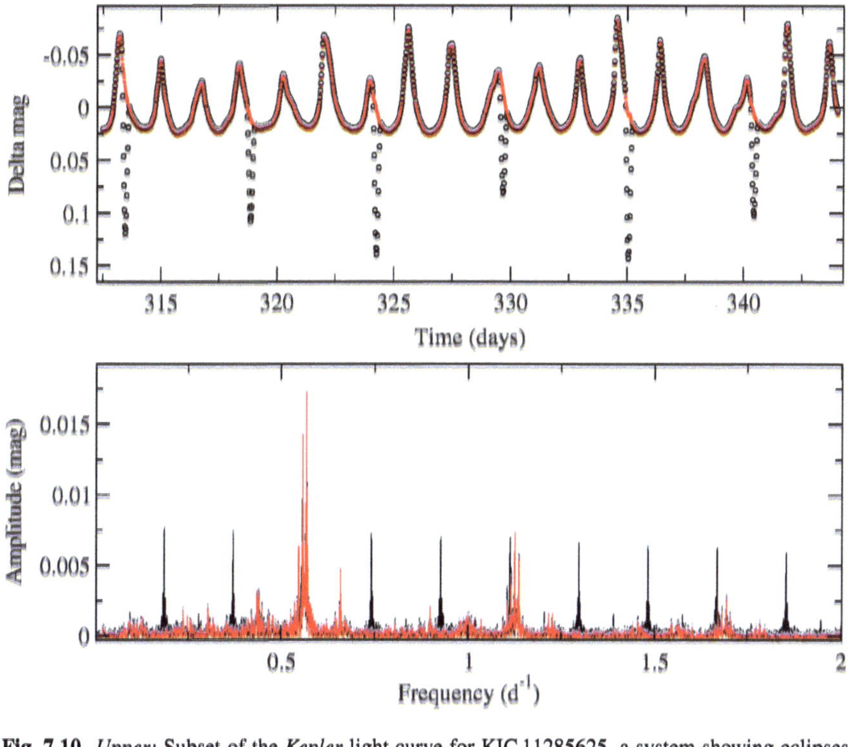

Fig. 7.10 *Upper:* Subset of the *Kepler* light curve for KIC 11285625, a system showing eclipses with a period of 10.8 days and γ Dor pulsations with periods of ~1–2 days. *Filled circles* indicate the light curve after removal of the eclipses. *Lower:* Amplitude spectrum of the *Kepler* data before (*black*) and after (*coloured/grey*) removing the eclipse signals (*Reproduced, with kind permission, from Debosscher et al. (2013)*)

Detailed asteroseismic modelling of these systems is complex. First efforts have concentrated on identifying the pulsating components by modelling the expected pulsation frequency ranges of p modes and g modes based on the system properties as constrained by the dynamics. For example, Maceroni et al. (2014) showed that the estimated mean g-mode period spacing measured in the F-star eclipsing binary KIC 3858884 is consistent only with pulsations in the secondary component, despite both having similar masses, and radii that differ by only $\sim$10%. Furthermore, it proved necessary to include convective core overshooting in the theoretical models in order to obtain agreement with the observations. The exceptional amount of complementary information in KIC 3858884 and in other systems promises to advance our understanding of pulsations in intermediate-mass stars in future modelling efforts.

In the case of more evolved stars, ground-based observations have yielded a handful of RR Lyrae and Cepheid pulsators in eclipsing binary systems (Pietrzyński et al. 2010; Soszyński et al. 2011). Systems like those are important for testing the masses derived from evolutionary models, and for measuring accurately the stellar distances through period-luminosity relations. As of writing, no detections of RR Lyrae or Cepheid pulsators in eclipsing binaries have been reported from analyses of space data, though that is probably related to the relative sparsity of such stars in the *Kepler* and *CoRoT* target lists.

7.5.2 Compact Pulsators

Asteroseismology of white dwarfs or subdwarf B stars enables one to address a wide variety of fundamental physics such as convection, crystallization, the properties of neutrinos, and the evolution on the extreme horizontal branch. Reviews include those by Winget and Kepler (2008) and Heber (2009). Discovering compact pulsators in eclipsing binary systems is extremely valuable, both for cross-checking the properties that have been derived by asteroseismology and for providing independent constraints that can improve seismic modelling.

Owing to the intrinsic faintness of white dwarfs and sdB stars, the detection of pulsations in eclipsing binaries that contain them is challenging. The benchmark system is PG1336-018, an eclipsing sdB + M dwarf binary with a period of 0.1 days (Kilkenny et al. 1998). The sdB component shows p-mode pulsations with frequencies ranging from $\sim$5000–7000 μHz; combining them with a full orbital solution demonstrated that asteroseismic modelling yields values for mass and radius that agree with the dynamical estimates to within 1% (Vučković et al. 2007; Van Grootel et al. 2013).

A spectacular second detection of a pulsating sdB star in an eclipsing binary has recently been revealed by *Kepler* data (Østensen et al. 2010). KIC 9472174 (2M1938+4603, $V \sim$12.3) has an orbital period of 0.12 days, with strong variations due to the reflection effect in the light curve (Fig 7.11). After a modelled light curve has been removed, the residuals show an unusually rich frequency spectrum

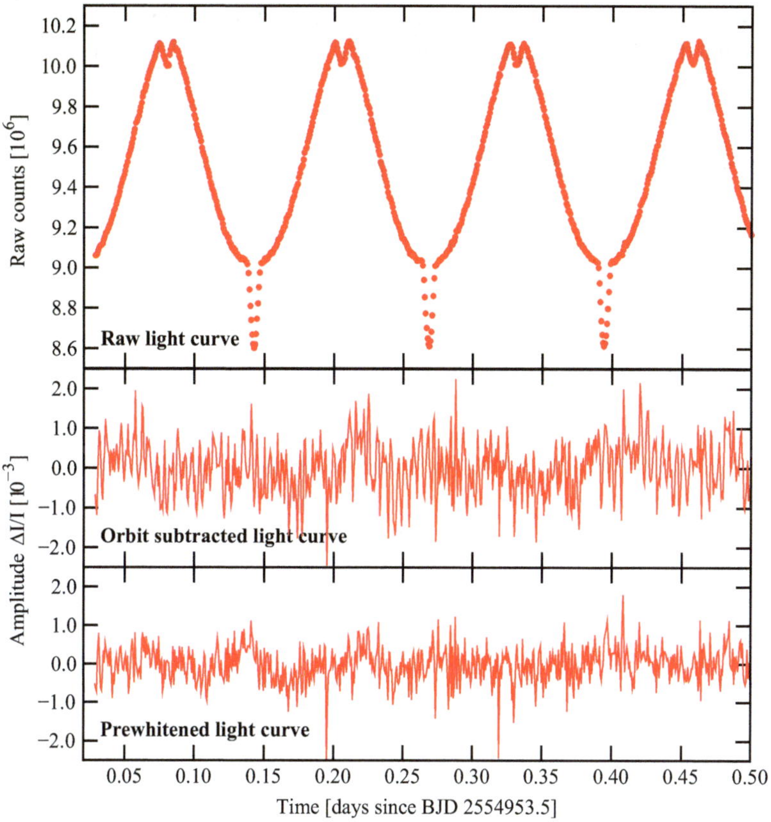

Fig. 7.11 *Upper: Kepler* light curve of the eclipsing sdB + M dwarf binary KIC 9472174. *Middle:* Light curve after subtracting the orbital solution. The pulsations in the sdB star are clearly visible. *Lower:* Residuals after subtracting modelled pulsation frequencies (*Reproduced, with kind permission, from Østensen et al. (2010)*)

in the sdB component. Østensen et al. (2010) report a total of 55 pulsation frequencies spanning from 50–4500 μHz, and attributed them to both p-mode and g-mode pulsations. The constraints from the orbital solution, combined with an RV semi-amplitude and spectroscopic gravity, yielded a mass estimate of $M = 0.48 \pm 0.03\,M_{\odot}$, which is consistent with a post common-envelope sdB star. Modelling the pulsation frequencies is expected to yield detailed constraints on the core structure, and hence on the progenitor mass of the sdB star.

Kepler has also uncovered several pulsating sdB stars in binaries that show light variations due to the reflection effect (Kawaler et al. 2010); they have been used to investigate tidal synchronization timescales by comparing rotation periods measured from rotational splittings to the binary period inferred from the light curve (Pablo et al. 2012). In addition, *Kepler* has uncovered an eclipsing (but non-pulsating) sdB + white dwarf binary which shows a combination of binary effects such as ellipsoidal deformation, Doppler beaming and microlensing (Bloemen et al. 2011).

7.6 Summary and Future Prospects

Table 7.2 summarizes the characteristics of confirmed eclipsing and/or heartbeat systems for which asteroseismic detections (either self-driven or tidally-induced) have been made with space-based observations. The list illustrates that the synergy of asteroseismology and eclipsing/eccentric binary stars using space-based observations is—perhaps not surprisingly—still in its infancy; all of the systems discussed have been published within the last 4 years. Most detections were made by *Kepler*, which provided not only the required continuous monitoring to detect eclipses, but also the high-precision photometry necessary to detect oscillations despite dilution by a binary component or the smallness of the amplitudes in the pulsating star. As mentioned in Sect. 7.4, the list of confirmed red giants in eclipsing systems can be expected to increase significantly once sufficient RVs have been gathered for the *Kepler* candidate systems (Gaulme et al. 2013).

Table 7.2 Eclipsing and/or heartbeat systems with components showing self-excited or tidally-induced pulsations detected from space-based observations. The systems are grouped into giants (*upper set*), intermediate and high-mass stars (*middle set*), and compact stars (*lower set*)

ID	V	Sp.Typ	P(d)	e	SB2	Stochastic	Coherent	Tidal	Ecl.	HB	References
HD 181068	8.0	KIII+MV+MV	45.5 + 0.9	0.0	No	No	No	Yes	Yes	No	c
KIC 5006817	10.9	KIII+MV	94.8	0.71	No	Yes	No	No	No	Yes	b
KIC 8410637	11.3	KIII+FV	408.3	0.69	Yes	Yes	No	No	Yes	No	a
KIC 10661783	9.5	G+AV	1.2	0.0	Yes	No	Yes	No	Yes	No	j
KIC 4544587	10.8	FV+FV	2.2	0.29	Yes	No	Yes	Yes	Yes	Yes	e
HD 174884	8.4	BV+BV	3.7	0.29	Yes	No	Maybe	Maybe	Yes	Yes	g
CID 102918586	11.7	FV+FV	4.4	0.25	Yes	No	Yes	Yes	Yes	Yes	h
KIC 11285625	10.1	FV+FV	10.8	0.0	Yes	No	Yes	No	Yes	No	i
KIC 3858884	9.3	FV+FV	26.0	0.47	yes	No	Yes	Maybe	Yes	No	f
HD 187091	8.4	AV+AV	41.8	0.83	Yes	No	No	Yes	No	Yes	d
KIC 9472174	12.3	sdB+MV	0.13	-	No	No	Yes	No	Yes	No	k

For each object (ID), the columns list approximate V magnitude, spectral type, orbital period (in days), eccentricity, and whether there is a double-lined spectroscopic orbit (SB2). The remaining five flags indicate the detection of stochastic oscillations, self-excited coherent pulsations, tidal pulsations, eclipses (Ecl.), and heartbeat effects (HB)

References: (**a**) Hekker et al. (2010) and Frandsen et al. (2013). (**b**) Beck et al. (2013). (**c**) Derekas et al. (2011), Borkovits et al. (2013), and Fuller et al. (2013). (**d**) Welsh et al. (2011). (**e**) Hambleton et al. (2013). (**f**) Maceroni et al. (2014). (**g**) Maceroni et al. (2009). (**h**) Maceroni et al. (2013). (**i**) Debosscher et al. (2013). (**j**) Southworth (2011) and Lehmann et al. (2013). (**k**) Østensen et al. (2010)

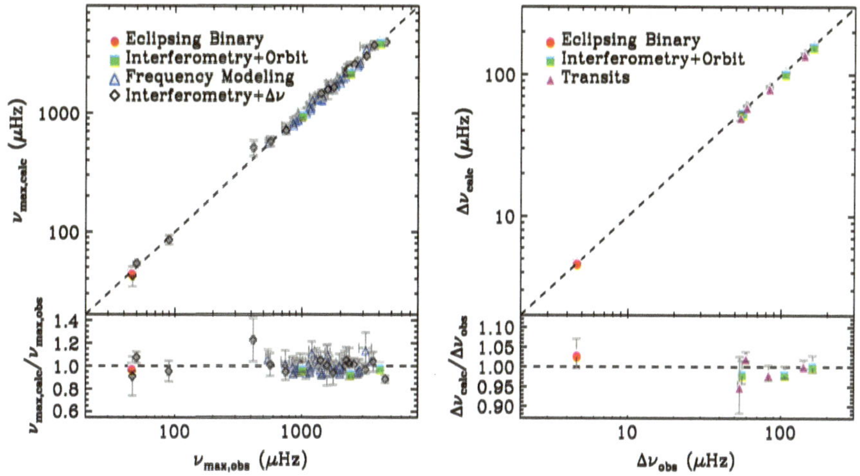

Fig. 7.12 Empirical tests of asteroseismic scaling relations. *Left:* Comparison of measured ν_{max} values with empirical values calculated by independent methods identified in the legend. *Filled symbols* are empirical values which are independent of asteroseismology. *Right:* As for the left plot, but for the $\Delta\nu$ scaling relation

Continued observations, and the search for new asteroseismic eclipsing binaries, remain crucial for answering important questions regarding stellar pulsations and evolution across the H–R diagram. To highlight one particular aspect, Fig. 7.12 shows updated empirical tests of the ν_{max} and $\Delta\nu$ scaling relations for stochastic oscillators. Note that most points in the ν_{max} comparison are not fully independent from asteroseismology, but have properties determined either by combining interferometric angular diameters with asteroseismic densities calculated from the $\Delta\nu$ scaling relation (e.g., Huber et al. 2012), or by including masses and radii determined from individual frequency modelling (e.g., Metcalfe et al. 2014). KIC 8410637 marks the first data-point from an eclipsing binary, and offers an independent test of ν_{max} and $\Delta\nu$ that is otherwise only possible for wide binaries with radii and masses measured from astrometry and interferometry. The $\Delta\nu$ comparison includes four stars which host exoplanets that have independently constrained eccentricities, thus allowing an independent measurement of the mean stellar density (Seager and Mallén-Ornelas 2003; Winn 2010); the four are HD17156 (Nutzman et al. 2011; Gilliland et al. 2011), TrES-2 (Barclay et al. 2012; Southworth 2011), HAT-P7 (Christensen-Dalsgaard et al. 2010; Southworth 2011), and Kepler-14 (Huber et al. 2013; Southworth 2012). Only stars whose calculated ν_{max} and $\Delta\nu$ have uncertainties smaller than 20% and 10% have been included in the comparison.

The median residuals for both quantities are close to zero, with a scatter of 7% for ν_{max} and 3% for $\Delta\nu$. While these numbers are encouraging, it is important to note that the observational uncertainties for ν_{max} and $\Delta\nu$ derived from *Kepler* data are typically up to a factor of 2 or more smaller (Chaplin et al. 2014). Moreover, comparisons for giant stars are effectively limited to one data-point, yet the vast majority of stars with asteroseismic detections are giants (see Fig. 7.3). Indeed, KIC 8410637 indicates that the ν_{max} scaling relation remains accurate for giant stars, while the $\Delta\nu$ scaling relation predicts a density which is too low by ~7%. Observations of stochastic oscillations in other eclipsing binary systems are needed to decide whether or not that offset is systematic, and to allow an *empirical* calibration of the scaling relations.

Future observations of asteroseismic eclipsing binaries that have giant and dwarf components can be expected from space-based missions such as K2 (Howell et al. 2014), TESS (Ricker et al. 2009) and PLATO (Rauer et al. 2013). What is particularly important is that those missions will observe stars which are significantly brighter than typical *CoRoT* and *Kepler* targets, thereby increasing the potential for independent constraints from ground-based observations such as long-baseline interferometry. There is little doubt that future space-based observations of asteroseismic eclipsing binary stars, combined with improved modelling efforts, will continue to play an important role in advancing the understanding of stellar evolution across the H–R diagram.

References

Aerts, C.: EAS Publications Series **64**, 323 (2013)
Aerts, C., Christensen-Dalsgaard, J., Kurtz, D.W.: Asteroseismology (Springer, Dodrecht, 2010)
Antoci, V., Handler, G., Campante, T.L., et al.: Nature **477**, 570 (2011)
Barclay, T., Huber, D., Rowe, J.F., et al.: ApJ **761**, 53 (2012)
Beck, P.G., Bedding, T.R., Mosser, B., et al.: Science **332**, 205 (2011)
Beck, P.G., Hambleton, K., Vos, J., et al.: A&A **564A**, 36 (2013)
Bedding, T.R.: ArXiv e-prints (arXiv:1107.1723, 2011)
Belkacem, K., Samadi, R., Goupil, M.-J., et al.: Science **324**, 1540 (2009)
Bedding, T.R., Kjeldsen, H., Campante, T.L., et al.: ApJ **713**, 935 (2010)
Bedding, T.R., Mosser, B., Huber, D., et al.: Nature **471**, 608 (2011)
Belkacem, K.: in Proc. Ann. Conf. French Soc. A. & Ap., ed. by S. Boissier, P. de Laverny, N. Nardetto, et al. (SF2A-2012), p. 173
Belkacem, K., Goupil, M.J., Dupret, M.A., et al.: A&A **530**, A142 (2011)
Bloemen, S., Marsh, T.R., Østensen, R.H., et al.: MNRAS **410**, 1787 (2011)
Borkovits, T., Derekas, A., Kiss, L.L., et al.: MNRAS **428**, 1656 (2013)
Bouchy, F., Carrier, F.: A&A **374**, L5 (2001)
Breger, M.: in Delta Scuti and Related Stars, ed. by M. Breger, M. Montgomery (ASPC **210**, San Francisco, 2000), p. 3
Breger, M., Balona, L., Lenz, P., et al.: MNRAS **414**, 1721 (2011)
Brogaard, K., VandenBerg, D.A., Bruntt, H., et al.: A&A **543**, A106 (2012)
Brown, T.M., Gilliland, R.L., Noyes, R.W., et al.: ApJ **368**, 599 (1991)
Brown, T.M., Latham, D.W., Everett, M.E., et al.: AJ **142**, 112 (2011)
Burkart, J., Quataert, E., Arras, P., et al.: MNRAS **421**, 983 (2012)
Carrier, F., Bouchy, F., Kienzle, F., et al.: A&A **378**, 142 (2001)
Carter, J.A., Fabrycky, D.C., Ragozzine, D., et al.: Science **331**, 562 (2011)
Casagrande, L., Silva Aguirre, V., Stello, D., et al.: ApJ **787**, 110 (2014)
Chaplin, W.J., Basu, S., Huber, D., et al.: ApJS **210**, 1 (2014)
Chaplin, W.J., Bedding, T.R., Bonanno, A., et al.: ApJ **732**, 5 (2011a)
Chaplin, W.J., Kjeldsen, H., Christensen-Dalsgaard, J., et al.: Science **332**, 213 (2011b)

Chaplin, W.J., Miglio, A.: ARA&A **51**, 353 (2013)
Christensen-Dalsgaard, J.: http://astro.phys.au.dk/~jcd/oscilnotes (2003)
Christensen-Dalsgaard, J., Kjeldsen, H., Brown, T.M., et al.: ApJ **713**, L164 (2010)
Christiansen, J.L., Derekas, A., Ashley, M.C.B., et al.: MNRAS **382**, 239 (2007)
Creevey, O.L., Metcalfe, T.S., Brown, T.M., et al.: ApJ **733**, 38 (2011)
Damiani, C., Maceroni, C., Cardini, D., et al.: Ap&SS **328**, 91 (2010)
De Ridder, J., Barban, C., Baudin, F., et al.: Nature **459**, 398 (2009)
Debosscher, J., Aerts, C., Tkachenko, A., et al.: A&A **556**, A56 (2013)
Derekas, A., Kiss, L.L., Borkovits, T., et al.: Science **332**, 216 (2011)
Dziembowski, W.A., Gough, D.O., Houdek, G., et al.: MNRAS **328**, 601 (2001)
Fabrycky, D.C., Ford, E.B., Steffen, J.H., et al.: ApJ **750**, 114 (2012)
Frandsen, S., Lehmann, H., Hekker, S., et al.: A&A **556**, A138 (2013)
Fuller, J., Derekas, A., Borkovits, T., et al.,: MNRAS **429**, 2425 (2013)
Fuller, J., Lai, D.: MNRAS **420**, 3126 (2012)
Gaulme, P., Jackiewicz, J., Appourchaux, T., et al.: ApJ **785**, 5 (2014)
Gaulme, P., McKeever, J., Rawls, M.L., et al.: ApJ **767**, 82 (2013)
Gilliland, R.L., Brown, T.M., Christensen-Dalsgaard, J., et al.: PASP **122**, 131 (2010)
Gilliland, R.L., McCullough, P.R., Nelan, E.P., et al.: ApJ **726**, 2 (2011)
Gizon, L., Solanki, S.K.: ApJ **589**, 1009 (2003)
Gough, D.O.: in Hydrodynamic and Magnetodynamic Problems in the Sun and Stars, ed. by Y. Osaki (Uni. of Tokyo Press, 1986), p. 117
Grigahcène, A., Antoci, V., Balona, L., et al.: ApJ **713**, L192 (2010)
Guenther, D.B., Kallinger, T., Reegen, P., et al.: Comm. in Asteroseismology **151**, 5 (2007)
Guzik, J.A., Kaye, A.B., Bradley, P.A., et al.: ApJ **542**, L57 (2000)
Hambleton, K.M., Kurtz, D.W., Prša, A., et al.: MNRAS **434**, 925 (2013)
Handler, G., Asteroseismology (Springer, Dordrecht, 2013), p. 207
Heber, U.: ARA&A **47**, 211 (2009)
Hekker, S., Debosscher, J., Huber, D., et al.: ApJ **713**, L187 (2010)
Hekker, S., Elsworth, Y., Basu, S., et al.: MNRAS **434**, 1668 (2013)
Hekker, S., Elsworth, Y., Mosser, B., et al.: A&A **544**, A90 (2012)
Hekker, S., Kallinger, T., Baudin, F., et al.: A&A **506**, 465 (2009)
Houdek, G., Balmforth, N.J., Christensen-Dalsgaard, J., et al.: A&A **351**, 582 (1999)
Howell, S.B., Sobeck, C., Haas, M., et al.: PASP **126**, 398 (2014)
Huber, D., Bedding, T.R., Stello, D., et al.: ApJ **743**, 143 (2011)
Huber, D., Chaplin, W.J., Christensen-Dalsgaard, J., et al.: ApJ **767**, 127 (2013)
Huber, D., Ireland, M.J., Bedding, T.R., et al.: ApJ **760**, 32 (2012)
Huber, D., Silva Aguirre, V., Matthews, J.M., et al.: ApJS **211**, 2 (2014)
Huber, D., Stello, D., Bedding, T.R., et al.: Comm. in Asteroseismology **160**, 74 (2009)
Kawaler, S.D., Reed, M.D., Østensen, R.H., et al.: MNRAS **409**, 1509 (2010)
Kilkenny, D., O'Donoghue, D., Koen, C., et al.: MNRAS **296**, 329 (1998)
Kippenhahn, R., Weigert, A.: Stellar Structure and Evolution (Springer, Berlin, 1994)
Kjeldsen, H., Bedding, T.R.: A&A **293**, 87 (1995)
Kjeldsen, H., Bedding, T.R., Butler, R.P., et al.: ApJ **635**, 1281 (2005)
Kurtz, D.W.: MNRAS **200**, 807 (1982)
Lai, D.: ApJ**490**, 847 (1997)
Lehmann, H., Southworth, J., Tkachenko, A., et al.: A&A **557**, A79 (2013)
Lehmann, H., Tkachenko, A., Semaan, T., et al.: A&A **526**, A124 (2011)
Maceroni, C., Lehmann, H., da Silva, R., et al.: A&A **563A**, 59 (2014)
Maceroni, C., Montalbán, J., Gandolfi, D., et al.: A&A **552**, A60 (2013)
Maceroni, C., Montalbán, J., Michel, E., et al.: A&A **508**, 1375 (2009)
Matijevič, G., Prša, A., Orosz, J.A., et al.: AJ **143**, 123 (2012)
Matthews, J.M.: Comm. in Asteroseismology **150**, 333 (2007)
Mészáros, S., Holtzman, J., García Pérez, A.E., et al.: AJ **146**, 133 (2013)
Metcalfe, T.S., Chaplin, W.J., Appourchaux, T., et al.: ApJ **748**, L10 (2012)

Metcalfe, T.S., Creevey, O.L., Dogan, G., et al.: ApJS **214**, 27 (2014)
Michel, E., Baglin, A.: arXiv:1202.1422 (2012)
Michel, E., Samadi, R., Baudin, F., et al.: A&A **495**, 979 (2009)
Miglio, A.: in Red Giants as Probes of the Structure and Evolution of the Milky Way, ed. by A. Miglio, J. Montalbán, A. Noels (Springer, Berlin, 2012)
Miglio, A., Brogaard, K., Stello, D., et al.: MNRAS **419**, 2077 (2012)
Miglio, A., Chiappini, C., Morel, T., et al.: Eur. Phys. Journal Web of Conferences **43** 3004 (2013)
Mkrtichian, D.E., Kusakin, A.V., Rodriguez, E., et al.: A&A**419**, 1015 (2004)
Mosser, B., Belkacem, K., Goupil, M.J., et al.: A&A **517**, A22 (2010)
Mosser, B., Michel, E., Belkacem, K., et al.: A&A **550**, A126 (2013)
Nemec, J.M., Cohen, J.G., Ripepi, V., et al.: ApJ **773**, 181 (2013)
Nutzman, P., Gilliland, R.L., McCullough, P.R., et al.: ApJ **726**, 3 (2011)
Østensen, R.H., Bloemen, S., Vučković,M., et al.: ApJ **736**, L39 (2011)
Østensen, R.H., Green, E.M., Bloemen, S., et al.: MNRAS **408**, L51 (2010)
Pablo, H., Kawaler, S.D., Reed, M.D., et al.: MNRAS **422**, 1343 (2012)
Pamyatnykh, A.A.: in Delta Scuti and Related Stars, ed. by M. Breger, M. Montgomery (ASPC **210**, San Francisco, 2000), p. 215
Pietrinferni, A., Cassisi, S., Salaris, M., et al.: ApJ **612**, 168 (2004)
Pietrzyński, G., Thompson, I.B., Gieren,W., et al.: Nature **468**, 542 (2010)
Prša, A., Batalha, N., Slawson, R.W., et al.: AJ **141**, 83 (2011)
Rauer, H., Catala, C., Aerts, C., et al.: arXiv:1310.0696 (2013)
Ricker, G.R., Latham, D.W., Vanderspek, R.K., et al.: BAAS **41**, No. 403.01 (2009)
Rodríguez, E., Breger, M.: A&A **366**, 178 (2001)
Salaris, M., Cassisi, S., Pietrinferni, A., et al.: ApJ **716**, 1241 (2010)
Seager, S., Mallén-Ornelas, G.: ApJ **585**, 1038 (2003)
Silva Aguirre, V., Casagrande, L., Basu, S., et al.: ApJ **757**, 99 (2012)
Slawson, R.W., Prša, A., Welsh, W.F., et al.: AJ **142**, 160 (2011)
Sokolovsky, K., Maceroni, C., Hareter, M., et al.: Comm. in Asteroseismology **161**, 55 (2010)
Soszyński, I., Dziembowski,W.A., Udalski, A., et al.: Acta Astron. **61**, 1 (2011)
Southworth, J.: MNRAS **417**, 2166 (2011)
Southworth, J.: MNRAS **426**, 1291 (2012)
Southworth, J., Zima, W., Aerts, C., et al.: MNRAS **414**, 2413 (2011)
Stello, D., Bruntt, H., Preston, H., et al.: ApJ **674**, L53 (2008)
Stello, D., Chaplin,W.J., Bruntt, H., et al.: ApJ **700**, 1589 (2009)
Stello, D., Huber, D., Bedding, T.R., et al.: ApJ **765**, L41 (2013)
Tassoul, M.: ApJS**43**, 469 (1980)
Thompson, S.E., Everett, M., Mullally, F., et al.: ApJ **753**, 86 (2012)
Ulrich, R.K.: ApJ **306**, L37 (1986)
Van Grootel, V., Charpinet, S., Brassard, P., et al.: A&A **553**, A97 (2013)
Vandakurov, Y.V.: Soviet Ast. **11**, 630 (1968)
Vučković, M., Aerts, C., Østensen, R., et al.: A&A **471**, 605 (2007)
Welsh, W.F., Orosz, J.A., Aerts, C., et al.: ApJS **197**, 4 (2011)
White, T.R., Bedding, T.R., Stello, D., et al.: ApJ **743**, 161 (2011)
Winget, D.E., Kepler, S.O.: ARA&A **46**, 157 (2008)
Winn, J.N.: arXiv:1001.2010 (2010)
Winn, J.N., Fabrycky, D., Albrecht, S., et al.: ApJ **718**, L145 (2010)
Zahn, J.P.: A&A **41**, 329 (1975)

Afterword

The primary aim of this volume has been to elucidate the nature of the outer atmospheres of cool giant and supergiant stars, particularly through observations and analyses of those immensely intriguing atmospheric eclipsing systems, the ζ Aur stars. In seven chapters, we have explored ground-based and space-based investigations of these objects, examined contemporary theory of chromospheric heating and stellar wind generation, and delved into recent determinations of the fundamental parameters that are needed to understand the structure and evolution of late-type stars. Along the way we took a beneficial foray into two interesting and complex systems, illustrating the power of new instrumentation and techniques since the last dedicated review of the subject of this book by K.O. Wright in 1970.

What new knowledge has been acquired, and—to be truthful—how much real progress has been made in relation to what still needs to be learned? How can we know when there is no more to be discovered? As is so often said, research opens up more questions than it actually answers, but would it be as satisfying if the reverse were the case? Rather, is it not a matter of moving the goal-posts so that there is always something new that needs investigating, new depths to plumb, new heights to scale?

Progress is never monotonic. Influenced by parallel developments in both theory and observation, and coupled with technological intricacies that enhance the way those activities can be combined and furthered, progress means moving away from the present state and into realms that are less familiar and require fresh learning-curves. We may not yet know very precisely what causes stellar oscillations, what controls turbulence in a stellar chromosphere, where a star's wind originates or what drives it, nor whether the radiation of the hot star (which alone brings to light all the fascinating details about a chromosphere which are specific to the subjects reviewed in these chapters) is a purely passive component or is actively modulating what is there through interactions at some level. The important thing is to know that those questions are *not* yet fully answered, that the physics *is* more complicated than (necessary) assumptions admit, that new theories do *not* explain everything,

T.B. Ake, E. Griffin (eds.), *Giants of Eclipse: The ζ Aurigae Stars and Other Binary Systems*, Astrophysics and Space Science Library 408,
DOI 10.1007/978-3-319-09198-3

and that the way forward is going to be challenging—to say the least. But though from some angles that situation may appear overwhelming and depressing, in other light it encourages persistence and strengthens the urge to enquire. Monitoring these stars with perseverance even with tested and tried techniques yields new facts that stimulate new ideas; applying new technologies provides new channels to inspire new information, and developing new theoretical approaches that build on state-of-the-art computing capabilities offers new ways to meld observation and interpretation.

All of the above is true of the small sample of ζ Aur stars which this book has attempted to describe. It is also true that the pattern of research into that genre of object has undergone changes and has had to meet challenges that have undermined, sometimes decimated, the resources that were once proportionately available. The progress described by K.O. Wright in his 1970 review enjoyed the benefits of a community whose majority was engaged in research into what today are termed bright stars; much equipment was optimised for it, and local groups flourished in the absence of serious competition. But restlessness at the inability to research what else was out there drove the development of electronic detectors that could observe much fainter objects, and computing technology (when it finally caught up) did the rest. Bright-star research lost its foreground priority, and appeared to be in danger of losing any priority at all, but the 'faithful few' who persisted through those dolldrum years now enjoy an unexpected re-emergence of substantial bright-star interest through the coincidental popularity of asteroseismology, planet-hunting and interferometric imaging. Add to those the new power of massive database mining skills and the unfathomed depths of citizen science, and we may revert once more to the challenge of having more observational material than the present workforce can cope with. Prising out more facts will not be easy—we will not be able to "see" what is going on in the tenuous outer layers of a huge evolved star as clearly as we could image what was actually happening during ingress into the recent eclipse of ϵ Aurigae—but we can be fairly certain that new observations will carry increasingly higher precision, and that will get us a long way towards improving the knowledge that we still seek.

Index of Keywords

T.B. Ake, E. Griffin (eds.), *Giants of Eclipse: The ζ Aurigae Stars and Other Binary Systems*, Astrophysics and Space Science Library 408,
DOI 10.1007/978-3-319-09198-3